SUCCESSFUL, WIN-WIN STRATEGIES FOR A SUPERYACHT PROJECT

WHAT MAKES OR BREAKS

THE CREATION OF A SUPERYACHT

TOMEK M. GLOWACKI

Successful, win-win strategies for a superyacht project
What makes or breaks the creation of a superyacht

Book cover design: Mikolaj Glowacki
Picture on book cover: Ken Freivokh Design

Tomasz (Tomek) M. Glowacki

Format: Soft cover
ISBN 978-0-473-30352-5

1st edition: April 2015

To my teachers, mentors, coaches, work colleagues and associates with whom together we achieved so much, and to all my contradictors, antagonists, adversaries and opponents without whom I would not challenge myself to achieve even more.

"IT IS LITERALLY TRUE THAT YOU CAN SUCCEED BEST AND QUICKER BY HELPING OTHERS TO SUCCEED"

-- NAPOLEON HILL

Figure 1 Captain's cabin, Grant Reed design

Author's Note:

Killing of lawyers referred to in this book is meant for literary and metaphorical purposes only. Such activities are immoral and illegal in most countries and we should restrain ourselves from murderous activities of all types, even under the strongest desire to commit such actions in certain circumstances.

What people say about this book

This AMAZING book should become compulsory reading for all clients, designers, project managers and yacht-builders. A true manual. The author has excelled with outstanding quotations by other notables." —**Alan Warwick, Yacht Designer, Auckland, New Zealand**

"Interesting and educational." —**Ted Fontaine, Yacht Designer, Newport USA**

"The book is clear and positive, a roadmap for success rather than a trip through misery lane. You need to read this book before you make your first phone call!" —**Marsha Cook Woodbury, PhD, Lecturer and Business Owner**

"Enjoyable, enlightening and impressing. If only it had been available years ago, then perhaps some owners and shipyards that I know would have benefitted from these words of wisdom instead of falling out and not getting the perfect 'win-win situation' that this book gives advice on achieving. It is required reading before starting a project, and then rereading during and after a project! A good piece of work based on a lifetime of knowledge and experience." —**Nick Gladwell, CEng, BSC(Hons), BA, FIMarEST, Former Director, Cayman Island Shipping Register**

"It would have been excellent to have this book in hand and thoroughly understood before starting our yacht. I would imagine that this book could be the source of a yardstick for both owner and builder so that it is somewhere to refer to when starting, as things progress through a build. This might also be something to think about for the ongoing relationship between the builder and the owner for the continuing opportunities and

satisfaction of both parties." —**Allen Jones, CEO of New Zealand Yachts**

"This book has a very good overall message in that building a superyacht, or any major project, should be a pleasant experience for everyone involved and the only way that can happen is for everyone to be 'preaching from the same hymn book'! Fairness is the essence of a successful project and even if an aggressive client thinks he has won by screwing a yard over, he has not. The negative energy flows from the management to the floor and no additional niceties are done to add to the finishing touches." —**Mark Stothard, Director of ECHO Yachts, WA**

"Very good, straightforward and concise book with valuable information that is easy and entertaining to read. More important, it is a valuable resource much needed in the industry. This book goes a long way to explain what people should think about when getting involved in a project." —**Mark Masciarotte, Marine Industry Consultant, Vancouver**

"I enjoyed reading this book very much. It shows a great understanding of the subject and I found myself nodding and smiling many times. It is a useful document." —**Richard Williams, Director at Big Blue Consultants Pty Ltd, WA**

"I have not seen anything written along such lines – it will certainly be of interest and a talking point in the industry, and the very easy, non-pompous, and relaxed style may appeal to lay people as well! Very good!" —**Ken Freivokh, designer**

"Tomek has hands-on, on-the-ground industry experience in planning and executing on superyacht projects like few others and is one of the few who can speak to this subject with any credibility. Before sinking any funds into any superyacht project, this book and a consultation with Tomek is a must." —**Michele**

Discepola, Lawyer at O'Melveny & Myers LLP, Singapore

"The success of a build or refit project is in the hands of the project manager and yet there is nothing else published that tells a project manager how to do his job properly. This book should be a must-read for yard managers, brokers, yacht managers and even designers because success is virtually assured if you put into practice all the wisdom within these pages." —**Butch Dalrymple-Smith, Naval Architect, France**

"This is a great book in that it clearly outlines the complete management process of any superyacht project. It demonstrates how common sense and good planning is essential to any successful project – how to start, how to proceed and how to finish. It offers practical advice and instruction and is easy to follow. It asks the reader to always consider 'what, why and how' in their planning processes. I think that anyone reading this book will instantly recognise where they may have in the past gone wrong and how to improve on their project management in the future; or, if they are new to it, then this book sets some great guidelines for how they should do it." —**David Chalmers, Director of IMED Ltd, Auckland**

CONTENTS

ACKNOWLEDGMENTS

This book would be even further away from perfection if not for the criticism, advice, comments, encouragements and help from: Bob Williams, Mark Stothard, Butch Darlymple-Smith, Ken Freivokh, Richard Williams, Bill Sanderson, Grant Reed, Nick Gladwell, Michele Discepola, Alan Warwick, Marsha Woodbury, Allen Jones, Michal Glowacki, Mark Masciarotte, Ted Fontaine, Robin Schroffel, Bronwen Morris, Ron Brown, Mikolaj Glowacki, and David Chalmers. Thank you all!

—Tomek

Successful, win-win strategies for a superyacht project

FOREWORD

I am taken by Napoleon Hill's quote, author of *The Law of Success in 16 Lessons*, for it is such a truth in life that in helping others we unmistakably help ourselves and this book is the perfect example of just that. In the belly of its creation is the rumble of self-evaluation, questioning, arguing with what should be here and what should not, what is important and what is not, what I know it to be and what I think it to be, the polishing and re-polishing until here it is – a compendium of knowledge and wisdom drawn from Tomek Glowacki's life's work in the world of shipbuilding.

As chairman and principal investor in New Zealand Yachts and the builder of m/y *Spirit*, winner of the Superyacht Society 2004 Superyacht of the Year 32-43m Power Class, I can attest that the journey of a yacht build can call into question one's sanity; however, I can add – and perhaps include those that sojourned with me might agree – that along with the yacht, one is called to build strength of character and tenacity.

The process of building *Spirit* required all that Tomek references in this delightful and long overdue dissertation on what to do, what not to do, what to watch out for and what to expect. It's all there in the cake mix of building a superyacht. This is the pocket book that by the end of the build will need to be replaced, for it will be tattered, written on, thrown against the wall and perhaps even jumped on, but it will carry those intending to embark on the journey of building a new superyacht through the storm out into calm seas and beyond.

—Allen Jones, LA, 5 February 2015

Figure 2 - m/y *Spirit,* Ken Freivokh Design

PREFACE

Choosing and building a superyacht should be as satisfying and challenging as any complicated undertaking, yet often the process takes too long, runs over budget, and results in an inferior boat that doesn't even meet specifications. In some cases, the brand new superyacht capsizes during the launching process, leaving a bad aftertaste of the whole experience for all involved. Why? How can such a massive project go wrong?

Over the past twenty-five years, ever since one of the first superyachts touched the water, we have been witnessing the development of an entire new sector of shipbuilding: the superyacht building industry. There are a number of superyachts being built around the world, so one may think that the process of building a superyacht is well advanced and each client/owner leaves the yard satisfied. This is far from the truth. Only a few yards can be classified as top performers. Superyacht building is still a relatively young industry. The International Superyacht Society was only founded in 1989; perhaps this is why the superyacht building process has still some hidden traps.

Except for a few glossy coffee-table books on superyachts, little has been written on this subject, particularly on the issues behind the scenes. So this book is my humble attempt to paint a picture of the superyacht building process as it stands today. By including short stories about court cases, anecdotes from the design office, and memorable construction moments, I hope this book will provide noteworthy, easily remembered information, tricks of the trade, and hints to help in the creation of superyachts. For the chapter's titles, I have deliberately used striking quotations (sometimes only with a symbolic meaning) so these chapters could be easily associated with their contents.

During my professional career I've had the privilege of working with exceptional people. I had wonderful mentors and was associated with world-class experts. Using my background and forty years of experience, visiting many yards all over the world and being lucky enough to work with extraordinary people, I will try to expose the atmosphere, difficulties, and hidden traps which accompany the construction of a superyacht. I've been involved in superyacht construction in various roles, from the floor through design office, and project manager to GM. Thanks to these different roles, I am able to vividly describe the weaknesses (and strengths) of the superyacht construction industry and give you information firsthand, straight from the gut, direct from the trenches. While auditing, or while on a casual visit to some shipyards, I could hear desperate and encrypted comments from the floor and cries for help:

"We were let down by other people."

"There is a constant lack of information on the floor."

"Changes, changes, changes…too many changes."

"Clients should not come to the workshop."

Do you see the frustration in these words?

This book is for everybody who was, is or will be, in some way or another, involved in building superyachts. It actually goes a bit further because it is also for anyone involved in any project creating a product, although those involved in maritime projects may find the issues a little closer to their hearts. This book is also about maintaining a healthy balance in business. Like balance in nature plays an important role in keeping all alive and flourishing, equilibrium in business plays the same role, where the competing influences of seller and buyer must be balanced. A win-win approach is so essential and is only possible when the yard's personnel and buyer are on the same page, so to speak. This book should help you to understand this philosophy.

Figure 3 - s/y *Savannah*, Ted Fontaine Design

INTRODUCTION

"A dream doesn't become reality through magic; it takes sweat, and determination. Success is the result of perfection, hard work and learning from failure."

These words were spoken by General Colin Powell in his presentation "A Leadership Primer." Building a superyacht is realising someone's dream, which requires careful planning, hard work and commitment. Things can go wrong, but to avoid this, plan for it. Use good, robust strategies and pay attention to detail, and if things go wrong anyway, strive for the win-win environment.

Simply speaking, building a superyacht is about three main strategies:
- Establishing an experienced and knowledgeable team that pays attention to detail.
- Operating in accordance with a detailed Quality Management System (QMS).
- Creating a win-win environment at all times.

In the following pages I'll be more specific and will break down these strategies into smaller portions, in this way highlighting that every project can be finished successfully, providing that we have knowledge and are able to adhere to the earlier established principles.

By talking explicitly about the definition of product, design and building processes, you will see why some yards perform better than others. Many difficulties faced by both parties are caused by a lack of skills and knowledge of project management methodology and a lack of awareness of how much and to what degree a company culture influences performance. In my experience, I've noticed that in some operations, people may know that something is going wrong but they have difficulty breaking problems down to smaller components and dealing with them. Sometime these are very small matters, such as a good review of the design or not enough clarity in a purchase order. But these small matters can make or break a project.

Here is an example of a small issue which could have grown to become a big problem, but was solved thanks to the win-win approach of both parties.

It is the end of the year 1998. I am the project manager for a 100-foot LOA (length over all) sloop in one of the most respected yards in the world. The design of the boat came from a well-known, very admired and accomplished design studio. The project is heading toward completion in a couple of months. This is the last client visit before the launching. The boat is well presented for this visit. The client, general manager, client representative, fit-out manager and I begin the inspection traditionally, from the cockpit. We then walk over to the mast area and the bow of the vessel. The client is impressed. Through the companionway, we are moving down below to the owner's cabin. It looks great! Then it is time to see the children's cabin. It's looking great too. But wait! Where is the porthole?

There was a porthole, but it was covered with joinery. Actually, you could look through, although you would have had to stick your head in a hole and squeeze your body upward. How could this have happened? Well, the designer wanted to keep all the portholes in one line when viewed from outside of the vessel. This collided with interior layout. By the way, such an occurrence is not an isolated case in superyacht construction.

The client was visibly disappointed: "Children need to have a porthole. They deserve to see the outside world. This has to be changed." Everybody looked at me.

Any changes at this point would cost a lot of money. Exterior fairing and painting and interior joinery was completed. To make changes now it would cost over $15,000 each porthole, and most probably an extension of construction for one month. The question was who was going to pay for it. I went through all the documentation. There were three drawings that showed the positions of these unfortunate portholes. All three drawings were approved by the client. Although I felt relief on my shoulders, it gave me little satisfaction because it didn't solve the problem. However, I felt in my gut that this problem would be solved somehow to the mutual satisfaction of both parties, and with this feeling I approached the meeting the next day.

We met early morning in the cabin again. I explained the work that had to be done and the cost involved in order to satisfy the requirement. For a moment we were at an impasse. Then the skipper (the client's representative) spoke: "Why don't we put in two mirrors, similar to a periscope?" he asked. It became quiet and everybody looked with anticipation at the client. After a moment of consideration he said, "This is a brilliant idea." Phew! Issue resolved! A solution was found because a solution had to be found, and came about because everybody wanted a solution. The next year, this yacht was the finalist for an international superyacht design award.

During the construction of a superyacht there are so many little things to consider. Even with the best team on board it is hard to avoid some mistakes or incoherencies. However, believing in teamwork and creating a win-win environment can help to overcome many issues.

Tomek M. Glowacki

General Description of the Industry

The sizes of yards in the superyacht industry range from a hundred to five hundred or more employees. Some yards (mostly European) have a long tradition in boat-building. Over their years of experience, they have developed sound management, strong leadership and good quality assurance programs. Their reputation is strong and they never suffer from a lack of orders. Others, for instance in New Zealand, were established around the mid-eighties and quickly gained a reputation. There are, however, a lot of newcomers. These are either totally new yards, or the yards which have been converted from either small boat-building or building of "black boats." These yards are struggling for reputation and have difficulties in convincing customers to place their order in their yard.

Some yards are more versatile than others. For instance, Alloy Yachts International builds exclusively from aluminium alloy, while the former Sensation Yachts used steel, aluminium and composite. Usually, the yard has its preferred size range, e.g. 80' to 120' or 100' to 150', etc. This is determined by the size of the existing facility and its launching ability.

The Characteristics of a Superyacht Project

This is how *ShowBoats International* magazine, in their January 2002 issue, defines a superyacht:

"Megayachts [initially they were called megayachts] originated approximately 10-15 years ago (1987 – 1992). It started with construction of several boats of around 100ft (30m) LOA. These types of yachts (power or sails) are typically custom built and feature very complex engineering and luxurious finishes. Those days a yacht over 80 feet (24.4m) was considered to be a

12

superyacht. Today they are considered superyacht if overall length is more than 100 feet although this is academic discussion."

This definition remains the same today. SUPERYACHT UK writes: "A super yacht is, by definition, any yacht over 24 metres (79 feet) in length – that's about the length of a tennis court."

By the way, the term megayacht was coined by the late Joseph Gribbins, who was the founder and editor of the publication *Nautical Quarterly*: "Megayacht was a boat that was so large that almost no one on earth could afford it".

There are many hundreds of superyachts sailing and visiting marinas, and they employ thousands of crew members from around the world. They can be motor or sail. Generally speaking, the ratio of sailing boats to powerboats in construction is 1 to 10. The largest superyachts in the world at the moment are ranging around 160-plus metres in length – that's larger than a Royal Navy Destroyer! However, the average length of a superyacht today is around 45 metres (148 feet).

Superyacht vs. Cargo Vessel

How different is the construction of a 40m LOA superyacht in comparison to a cargo vessel of 175m LOA? Just to give you some idea:

The displacement of a 40m LOA superyacht could be in the range of approximately 140 tonnes. It takes 20 months to build and consumes approximately 150,000 man-hours. This translates to 7t/month, and about 1100 man-hours/tonne. The displacement of a 175m LOA cargo vessel may be in the range of approximately 7500 tonnes. It takes nine months to build and consumes approximately 650,000 man-hours. This translates to

836t/month and 85 man-hours/tonne. From the data above you can see how the construction of a superyacht is more labour-intensive in comparison with a much larger cargo vessel – almost thirteen times more labour-intensive, in fact.

Each superyacht is unique; it is a one-off project. She is built of custom-made components. There is a high level of sophistication. Joinery is handmade and represents art more than trade. Sophisticated lighting provides a different mood for different occasions. Electronic equipment includes excellent entertaining, navigation, communication and security systems. There is a high level of luxury, sometimes even including gold-plated ironmongery, silver cutlery, and genuine artwork supplements in the interior.

Other characteristics of a superyacht project include (although some yards may not see it that way):

- Always high expectations, but often poor initial definition of product.
- Design specifications mixed with performance specifications.
- Creation follows the naval architectural "design spiral," which means a long and iterating process.
- High grade, high quality, and high complexity, resulting in high cost.
- Expensive art items.
- A lot of innovation.
- Small working space for a lot of people, particularly during the final fit-out.
- Required 100 percent reliability first-time. ("3.4 defects per million opportunities" not acceptable. Six Sigma Plus is not good enough.)
- Often a new technical experience for yard and client.

- If a yard is born on the foundation of a small boat-building business, it is limited in experience and knowledge.
- If there is lack of project management skills it leads to overlapping responsibilities and ambiguous roles of stakeholders.
- Non-commercial nature of project results in slower decision-making.
- Usually multi-project environment (if yard runs two or three boats at the same time, there could be competition for resources).
- Constant changing of critical path (in fact, always several critical paths at once).
- Expected (but not always delivered) fun aspect.
- Involvement of many parties (multicultural, international teams).
- Extensive documentation.
- Big egos? No, not really!
- Changes in scope of work and technical specifications. Casually implemented changes that cause disagreement.
- High-tech product and services create high risk.

Does it mean that if you have all of the above components and characteristics, you are dealing with a superyacht project? No, not necessarily. It is as difficult to describe a superyacht as to describe a true lady, but when you see her, you know it.

Buyers of superyachts of a value between $10-30m or more have a lot of enthusiasm, faith and courage, but they also need knowledge, skills and experience. The project of constructing a superyacht needs the efforts of both parties – shipyard and client – in order to perform well. When building a superyacht, both parties only have one shot. Paolo Scanu, the late, great Italian naval architect, used to say, "If this was an airplane, 500 designers would spend five years and 50 million dollars making 50 prototypes before they build it. With the superyacht we have to get it right first time. We have one shot."

Customers

People who are associated with boats often have a fondness or even affection for them. This sometimes is exposed in near-poetic descriptions of the features and the beauty or the process of creating them. Here is an example of such a metaphoric comparison by Jack A. Somer (excerpt from *Juliet: The Creation of a Masterpiece)*:

"The making of a sailing yacht is an intense synergy between art and science, between heart and hand: It is a striking harmony between rough sketches on foolscap and crisp printouts on Mylar; a translation of vague mental conception into hard-edged lofting; a focusing of diffuse artistry into plasma-sharp cutting; the landing of euphoric flights of imagination by a micro chip's swift calculation."

Each customer is unique and is in the game for a slightly different reason, and each is trying to obtain something special from his or her project. Despite the vast differences in personalities, there are, in general, only two types of customers:

One is a yachtsman, the client who commissions a new-build because of his love for boating. Although these clients might be the most opinionated due to their experience, they are, more often than not, the easiest customers to please. The reason is that their love for boats is genuine, and most of them enter into a boat-building project with more than a basic knowledge of the process and a sincere desire to accomplish something better or different than what has been done before. Usually, they are commissioning bigger boats than the last one. There is a saying that the waterline length of the customer's next boat represents the overall length of the former one.

Then there are clients who build boats for entirely different reasons. These individuals might indeed enjoy owning and being aboard a yacht, but they are hardly yachtsmen in the traditional sense of the word. Their motivation is different, their goal more superficial. What they want from their new boat is to make a statement about themselves, by which to increase their prestige.

One dynamic that comes into play with either of these types of clients is that during the course of the yacht-building project, a subtle but significant role change takes place. The client, who is accustomed to being a leader in his everyday life, suddenly finds himself surrounded by experts who know far more than he about the subject. Stranded outside his or her field of expertise, he or she can be a participant in the process, but not the leader. For the client, this can be an uncomfortable position.

Why Is It So Hard?

The creation of a superyacht is influenced by many stakeholders and the way they do business: the owner and his family, the future skipper and crew, the broker, the naval architect, the exterior stylist, the interior designer, suppliers of various equipment and materials, the classification society, the owner representative, the shipyard's project manager, various subcontractors, and finally, the yard itself. This is a large team of contributors that can make the project very successful or utterly disastrous – in other words, they can make it or break it. The construction of a superyacht requires a very high standard of theoretical correctness and a high level of practical experience. There are many processes to be managed, such as definition of scope of work, design, planning, procurements, etc. The chain of events is long and complicated, and many things can go awry – but they don't have to.

People who commission the construction of superyachts are usually well accustomed to business matters. They have probably signed a number of multimillion-dollar contracts, participated in the preparation of scope of works, specifications, and changes during production procurement, risk, etc. However, superyacht construction is a special kettle of fish. Clients want only three things: excellent quality, completion on time, and costs within the budget.

So, why is it so difficult for some to achieve those three things? The short answer is: the culture of and synergy between yard and client. It is also the attention to detail (or lack of it) which makes the difference. But which details? ALL OF THEM! To ensure success in a superyacht project, every stone must be turned.

Why Projects Fail (in General)

A NASA study of 650 projects (Kepner–Tregoe) shows the following reasons for project failures:

- Starting without a clear, defined project goal
- Wrong project manager
- Lack of team participation
- Lack of upper management support
- Improper structuring and use of project management process
- Unrealistic and incomplete project schedule
- Reluctance to end the project
- Lack of expertise

OK, so this is from NASA. But, are these reasons relevant to the superyacht industry? In my experience, they are very relevant and, in fact, they are relevant for any industry. They are the reason for the contents of this book. However, I would like to add one more reason specific to superyacht construction. This is due to the fact that documentation is supplied to the yard by the naval architect, who has been employed by a client. Sometimes there is a lack of understanding by inexperienced shipyards that the documentation received from a stylist/naval architect is not complete and still requires a lot of work by the in-house design team. In other words, it is confusion about responsibilities. I have seen this on a number of occasions. Let me expand on this.

On one occasion, a client spent an initial $600,000 on a concept drawing, external and internal styling. Then he spent another $600,000 on naval architectural design (structure, hydrostatics, stability, etc.). When he went with these drawings to a shipyard, they explained to him that they would have to spend about another $600,000 or more for detailed construction drawings, and shop drawings. The client could not understand and he was very upset. He was under the misconception that the documentation from the stylist and naval architect was all that needed to be done. Often, inexperienced shipyards are under the same illusion.

I've visited a number of shipyards and several of them did not have a design team at all. On one occasion a boatyard in the Middle East, which had plans to become a superyacht builder, had only one designer – or I should say, draftsman. When I asked why they did not have more designers to speed up the process, the answer was that the people in the factory could not read the drawings. Well, that was an extreme case of a company that had the ambition to build superyachts. So far they have not achieved their goal. On the other hand, I have seen other superyacht construction facilities where the design office has up to thirty designers or more. And, the yard with big design office usually perform very well.

Owners often pay a substantial amount of money for initial documentation, and it is hard for them to accept that the design is not complete and still requires between ten to twenty thousand or even more design hours.

It is worth mentioning here that responsibility for performance rests with the boatyard, even though the designer may have made some decisions that impact on it. One of the reasons for this is that a designer typically has very few assets. But the boatyard is a fixed establishment that owns buildings, equipment, materials, and goodwill. Lawyers do not attack people unless they have the wherewithal to meet a possible claim for damages.

What Is Strategy, Anyway?

The principles of modern strategy took shape with the growth of professional armies and the Napoleonic campaigns of the early part of the nineteenth century. Although the word "strategy" comes mainly from military language, today it is widely used in many other disciplines, such as business, marketing, thinking, operation, planning, various sports, games, etc.

Strategy is the science and art of a high-level plan of employing all resources to achieve goals under conditions of uncertainty. Strategy has many definitions, but generally involves setting goals and determining actions to achieve these goals by mobilizing resources.

Strategy has traditionally been distinguished from tactics in the following ways:

"Strategy deals with the entire theatre of war and the use of battles to win the war, while tactics is concerned primarily with the use of troops and equipment to win the battles. Strategy deals with the favourable positioning of troops as a prelude to battle while tactics is concerned with the handling of troops on the battlefield." —Encyclopaedia Britannica.

I already mentioned three main strategies which should guide superyacht construction projects, and I will dwell on them shortly. But there are many more little ones. Let's call them habits. One of my habits when managing the construction of a superyacht project is to review the construction schedule religiously every day for a minimum of fifteen minutes. I notice that initially, for the first five or ten minutes, I see nothing interesting, because I'd reviewed and updated it the day before. But after a while the issues start emerging. Staying positive, I constantly ask myself what could go wrong. Another one of my habits is to deal with the problem as soon as I hear about it.

It is beyond the scope of this book to explain all the tricks of the trade, processes and methodologies. So, I will concentrate only on those disciplines and strategies I believe are usually overlooked or neglected, but have a direct influence on the outcome of a superyacht construction project.

Something to remember:

Tactics come and go, but strategies are forever

*   *   *

21

SUCCESFUL, WIN-WIN STRATEGIES

FOR A SUPERYACHT PROJECTS

WHAT MAKES OR BREAKS
THE CREATION OF A SUPERYACHT

Before I get going, please let me allude to one thing: a superyacht is built by a shipyard, but the client also has responsibilities. Everything in this book relates to, and is important to, both parties: the shipyard and the client. To create a win-win environment, both parties to the contract have to work hard towards this goal.

Figure 4 - From the drawing board of Sam Sorgiovani and One 2 Three Naval Architects, to be completed in 2017 by Echo Yachts

Tomek M. Glowacki

STRATEGY ONE

Guide your legal team wisely

"FIRST, KILL ALL THE LAWYERS"

Few people are unfamiliar with the phrase "The first thing we do, let's kill all the lawyers." It has been strongly said, but this is Shakespeare talking in *Henry VI, Part 2.*

"This phrase often expresses the ordinary person's frustration with the arcana and complexity of law. Contrary to popular belief, the proposal was not designed to restore sanity to commercial life. Rather, it was intended to eliminate those who might stand in the way of a contemplated revolution – thus underscoring the important role that lawyers can play in society".
—Seth Finkelstein

Don't get me wrong – we need lawyers. Usually we need them at the beginning of the project when preparing a contract, and at the end of project if things don't go according to "wishful thinking". Often things go wrong if the contract has not been prepared correctly, and preparing a contract correctly is not only the lawyers' responsibility. The contract is a document we use to "marry" a client and a shipyard for approximately two years. Understanding a contract is the responsibility of all involved. I often hear, "This is lawyer's talk; I don't have to understand it." With such an approach, we are voluntarily depriving ourselves of the right to have a voice. This is disempowering! Both parties should understand every word in this "lawyer's talk," a contract.

Allow me to tell you a story about reviewing Ronald Reagan's slipway (repair yard) in Pago-Pago, American Samoa.

The governor of Pago-Pago and local businesses were concerned about the dilapidated state of the aforementioned slipway. They invited a group of consultants to review the situation and come up with a solution on how to improve/rectify it.

I went there with two consultants, one specialising in slipways operations and the other an expert in environmental matters. The yard was operated predominantly by retired American Navy personnel and, as we later discovered, they were more interested in fishing, playing golf and having a good life than the maintenance of Purse Seiners.

Before we went to the yard, we visited the governor and asked for the lease contract. It was a thick document and took me a few hours to read. I finished it at 5 a.m. the next day. However, within the first few pages, I realised that this contract was written in such a way that the governor could do nothing until the end of the lease, and even then the leasing company had first rights to re-lease the yard. No clauses about taking over in case of bad management were present in this document.

Why did nobody signing this document some years ago on behalf of the Government of Western Samoa see this coming? Was it just another case of "lawyer's talk"?

Whenever someone passes a contract to you for reading, consider this: If you don't understand all the contract's documents, you are not a part of the solution. If you are not a part of the solution, there is good money to be made in prolonging the problem. Someone, sooner or later, will take advantage of it. There is something else with regards to legal services: Each party to a contract has its own lawyers. They don't work together, they work against each other, and... they have big egos. Believing that other party is going to be kind to us is utopia.

This reminds me of a situation when I was employed to improve the way in which one of the New Zealand government's engineering groups managed their various projects. The management of one of the groups was attempting to sign two long-term contracts for the delivery of certain goods. A so-called "food fight" between the government's lawyers and two other companies' lawyers had already lasted for two years. There was always enough complexity to keep the argument going.

It wasn't only the lawyers' fault. The jobs had been given to them without exact explanation of what they were all about. The worst thing was that nobody had told them what wasn't in their scope of work. The scope of work, specifications and an agreement were all mixed up together in one document, and each party was trying to cover its back unrelentingly. Unfortunately, such cases are not isolated.

The solution to this problem was simple. All work for the preparation of the contract's documents was divided and distributed to the appropriate people: technical matters (specifications) to the construction manager, commercial matters to the accountant and scope of work to the project manager. When the job was completed, it was presented to the lawyers with a request to prepare the legal part: the contract agreement. The result was such that both contracts were signed within a couple of weeks, and celebrated with champagne.

Things vary between countries. The rule of thumb for a written contract is that the same deal will require a handshake in Poland, a one-page contract in Scandinavian countries, four pages in England and sixteen pages in USA. Guess where you're most likely to go to court?

This type of battle between lawyers is common because they often have big egos and try hard to establish their authority. Jack Welch, former CEO of General Electric understood this very well. In the book *Straight from the Gut*, he says, "We also agreed to control the egos of lawyers and bankers. Those outside teams often engage in food fights to prove who's smarter. I told Norm, 'Whenever that starts, let's get on the phone and resolve it quickly.'"

Benjamin Franklin had a similar view: "A countryman between two lawyers is like a fish between two cats."

I am not saying to get rid of lawyers altogether. You need lawyers, but first of all, you need to know what YOU want to achieve. Too often lawyers are given the task of writing a contract when it is only vaguely explained what the contract is all about. They start guessing and, in extreme cases, they are putting everything in one document: the contract agreement, scope of work, specifications, etc.

My 8 Ways to Keep Lawyers at Bay and Create a Win-Win Scenario:

1. Build your credibility with your knowledge and experience.

The first thing the lawyer of an opposite side tries to do when you enter the witness stand in a courtroom as an expert witness is destroy your credibility. They do it on purpose in order to diminish the value of everything you are going to say. They will start asking you about your years spent in the industry, your experience in a particular subject, your education, etc. Be sure to prepare yourself. The key to success is to learn as much as you can because when you are not learning, someone, somewhere else is.

2. Always do your homework

Always be well prepared for a talk with your partner (regardless of whether you are the client or yard's owner). Knowing all documents inside out is a must. Knowing all procurements and the delivery date of all items is a must. Knowing the construction schedule is a must... the list goes on. Here is the endorsement from Winston Churchill:
"To every man there comes...that special moment when he will be figuratively tapped on the shoulder and offered the chance to do a special thing unique to him...What a tragedy if that moment finds him unprepared or unqualified for that work which could have been his finest hour."

3. Keep everything well documented

By having all papers well filed and organised, you gain internal peace and the respect of others. Having in-depth knowledge of all pertinent information empowers you and gives you confidence. Nothing is worse than when you are asked for some documents and you cannot find them. Relying on just memory may, sooner or later, prove fatal.
"Blunt pencil will always remember more than a sharp mind!"
—Anon.

4. Maintain moral superiority with strong integrity

Moral superiority is an inner strength, and inner strength you can only have with quiet consciousness and strong integrity.

5. Don't get emotional, stay professional

It can be frustrating when things don't go your way, but if you let people push your buttons easily, your actions will jeopardise you. The ability to express and control our own emotions is important, and even more important is our ability to understand, interpret, and respond to the emotions of others. We can't control the world changing around us, but we can control how we respond to it. Some experts even suggest that emotional intelligence (EI) can be more important than intelligence quotient (IQ). So, instead of getting emotional, angry or frustrated, have a professional and pragmatic discussion. Keep this in mind:

"Anybody can become angry – that is easy, but to be angry with the right person and to the right degree and at the right time and for the right purpose, and in the right way – that is not within everybody's power and is not easy." —Aristotle (384-322 BC)

6. Be proactive

Putting it more bluntly, offence is the best strategy for defence. It is much easier to write a letter first and put everything into your own words than to get a letter from Mr. XYZ and then try to untangle everything.

"Both parties to a contract must be active participants during performance; passive contract management is taxed, active contract management is rewarded." —K.W. Fisher, Fisher Maritime

7. Settle disputes in the conference room, not in the courtroom

You cannot ever enter the courtroom with full confidence. Even if you think you have all the cards, you may be proven wrong.

Why? This may be due to an unforeseen development or because in court, events can sometimes take an awkward turn.

Before you file a lawsuit in court, you should always consider whether you can resolve your dispute out of court. By using Alternative Dispute Resolution (ADR) to resolve your disputes without going to court, you can:

- Save a lot of time
- Save a lot of money, and
- Have more control over the outcome.

Of course, sometimes it is impossible to avoid court. Maybe you are being sued for "no reason" and you just have to defend yourself or your company. For instance, you make a product on time for Ex-Works delivery. Your client has arranged transport but due to a problem caused by a cyclone, the shipment arrives two weeks later. The client sues you for late delivery. This is a real-life example.

8. Don't show your hand

Sun Tzu (544 BC–496BC), a Chinese military general, strategist and philosopher, believed that the key to victory was to get the enemy to reveal as much as possible while keeping him in the dark about your plans, abilities and knowledge. Well, this is not rocket science, but it is a good reminder of everyday strategies. It is not about hiding, either.

To conclude the above discussion, or rather, divagation on contracts and lawyers, here is a concentrated dose of wisdom from the very reputable Dr. Kenneth W. Fisher of Fisher Maritime: "Contracts developed by lawyers ensure availability of post-delivery remedies; but one developed primarily by contract managers reduces the need for post-delivery remedies."

Something to remember:

A lawyer may prepare a contract, but you have to sign it.

Figure 5 - s/y *Georgia*, Glade Johnson, Butch Dalrymple-Smith, and late Paolo Scanu design

STRATEGY TWO

Use Proven Project Management Methodology

BEFORE WE SET SAILS: LEARN THE BASICS

Business study, project management, planning, and leadership are common-sense subjects, although common sense is not always common. Experience and insider knowledge play a big part in building a superyacht. So, in the next few paragraphs I will touch on some basics related to project management.

The easiest way to grasp the concept of superyacht project management is to see how the five main project management processes overlap. See the figure below.

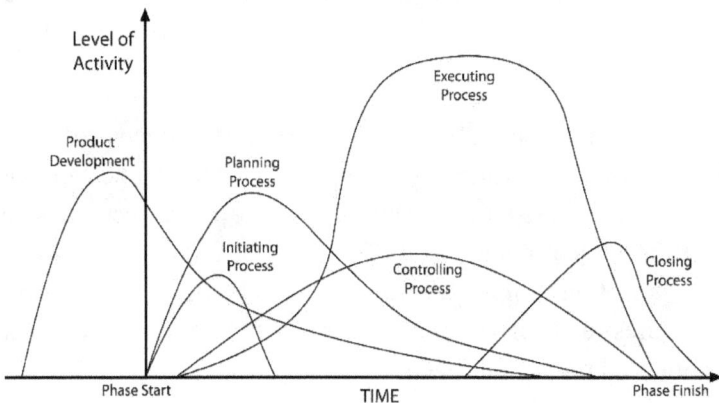

Figure 6 - Overlapping of Product Development Process with Project Management Processes in Superyacht Project

The creation of a superyacht overlaps two processes: product development and project management. In the process of development of products such as vessels, aircraft, tanks and similar, the design spiral process as described in Strategy 6 plays a big part.

At this point it should be emphasized that efficiency of the product development process can only be achieved if the requirements are clearly defined. According to the Project Management Institute, "inaccurate requirements" is the primary reason for 37 percent of projects failing. In other words, if you don't know where you are going, you will never get there.

Project Management Processes

In today's business, project management has becomes one of the most important methods for reducing risk, costs and improving success rate. Ninety percent of global senior executives ranked project management methods as either critical or somewhat important to their ability to deliver successful projects and remain competitive, according to an Economist Intelligence Unit survey.

According to the book *A Guide to the Project Management Body of Knowledge*, "Project management is the application of knowledge, skills, tools and techniques to project activities to meet the project requirements.". So one of the most important strategies of any shipyard is to set up good project management procedures and bring to the yard the best and the most experienced project managers.

The project management process consists of definition, planning, execution (monitoring and controlling) and closing. Sometimes it is also presented as in the figure below:

```
        ┌──────────┐
        │  DEFINE  │
        └────┬─────┘
             ↓
        ┌──────────┐
        │   PLAN   │
        └────┬─────┘
             ↓
        ┌──────────┐
 ┌─────→│ EXECUTE  │──────┐
 │      └────┬─────┘      ↓
┌┴───────┐   │      ┌──────────┐
│ ADJUST │←──┼──────│ MONITOR  │
└────────┘   │      └──────────┘
             ↓
        ┌──────────┐
        │ COMPLETE │
        └────┬─────┘
             ↓
        ┌──────────┐
        │ EVALUATE │
        └────┬─────┘
             ↓
        ┌──────────┐
        │CELEBRATE │
        └──────────┘
```

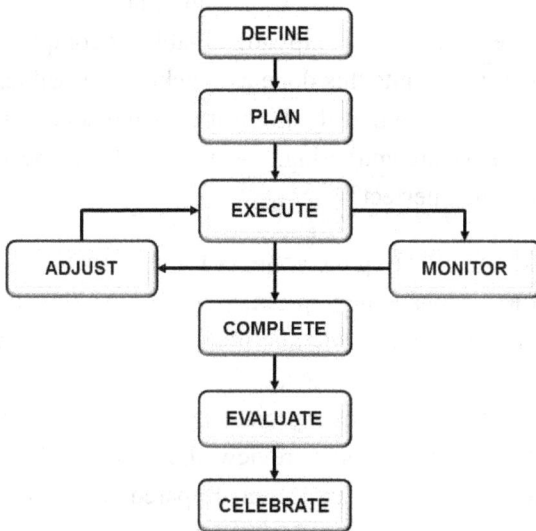

Figure 7 - Project management processes

The Initiating Process

Nobody will remember how we started the project, but everybody will remember how we finished. However, the first 5 percent of time spent on a project has 65 percent of the influence on the final result. This is why the initiating process of a project is so important! As Plato said, "Beginning is the most important part of any work." However, Napoleon Hill is right too: "Most of us are good starters but poor finishers". The question is, are we the wise starters? Do we know how to initiate a project?

A superyacht project is initiated by a client (and his entourage) by discussing the concept of a vessel with a naval architect and/or stylist. Then the naval architect, equipped with a design brief from the client, will begin sketching the client's future vessel. When he completes the preliminary design (this may take a long period of time and undergo many changes) the client's team begins searching for a shipyard capable of completing this project. Usually he wants this done as quickly as possible, for as little cost as possible and at the highest possible standard. Now, shipyards' owners are muttering, "Aren't we lucky he doesn't want now, free and perfect?"

The shipyard, after signing the contract for construction, initiates its own project management process. Assuming that the yard already has its own management system in place, this is nothing more than authorizing the project – the general manager assigns a project manager and issues a project charter. A project team is created and its first role is to review the contract documents. Because these documents are often prepared in a hurry, they need to be carefully scrutinised after signing the contract. This phase must begin with a thorough review of all client and vendor documents such as the contract, specifications, drawings and any other document attached to the construction contract. Delivery dates and other contractual obligations should immediately be put into a construction schedule as milestones. As a result of this study, a comprehensive list of things to do (particularly those immediately impacting on hull/deck construction) should be prepared. Taken into account should be such things as missing details, things to further review, a list of shop drawings, urgent designs to be completed, etc. It may take at least a week for a couple of very experienced people to perform the above tasks, but it is time well spent and it is absolutely necessary. After such a review, the following tasks usually require immediate attention. I call them "urgent designs":

- Geometry and structure
- Deck layout
- Penetrations (these are associated with reticulation planes)

- Through hull fittings
- Drive train arrangement
- Thrusters
- Stabilizer
- Anchor system
- Stern platform
- Doors and closures
- Windscreen
- Life-raft stowage and deployment
- Air in and out of the vessel
- Passerelle and gangway
- Fire protection plan
- Navigation lights positions
- Installation of inflatables

The detailed design and position of the above details has an impact on the hull and deck fabrication and, if not finalised early enough in the design process, may cause a delay in completion. I know! This is so obvious.

Something to remember:

Define, review, scrutinise, check, and check again. Measure twice, cut once!

The Planning Process

Planning is the process of thinking about and organizing the activities required to achieve a desired goal. The amount of planning is, and should be, proportional to the scope of the project.

After "the initiation," the planning is the most important process. It starts immediately after the signing of the contract and is an ongoing effort throughout the project's life. I would like to stress here that planning consists not simply of preparation of a project schedule and finding a critical path. It is quite a complex process and here I only bring attention to the core processes of planning:

- Scope planning
- Scope definition
- Activities' definition
- Activities' sequencing
- Activities' duration
- Schedule development
- Risk-management planning
- Resource planning
- Cost estimating (pre-contract)
- Cost budgeting
- Project plan development

A project plan is often confused with a project schedule. A project plan is a document or collection of documents, summarising guidelines and assumptions for project execution and project control. A project schedule is a part of the project plan and presents activity sequences, duration and resource requirements.

Something to remember:

Don't start a day without finishing it on paper.

The Executing Process

The executing process is putting the project plan into effect. This process includes:

- Project plan execution
- Quality Assurance
- Team development
- Information distribution
- Procurement, and
- Administration.

Monitoring and Controlling Process

The key benefit of this process is that project performance is observed and measured to identify any variances from the project plan. The monitoring and controlling process includes the following project management processes:

- Monitoring and controlling project work
- Change control
- Scope verification
- Scope control
- Schedule control
- Costs control
- Quality control
- Reporting
- Risk management
- Procurement contract(s) control

As I mentioned already, it is important that both parties to the contract take an active role in controlling the project.

To help us to monitor how we perform we use a trend analysis –

earned value. Earned value analysis (S-curve) tells us what we actually accomplished and estimates what will happen to the project in the future.

Part of an Execution Process Is Earned Value – Trend Analysis

Earned value analysis (S-curve) tells us what we actually accomplished and estimates what will happen to the project in the future. These analyses are rarely seen during superyacht construction and this is why: to gain the full benefit of this analysis, they have to be done properly and the process starts much earlier, during creation of Work Breakdown Structure (WBS). Here is a very important tip, often neglected or completely overlooked.

In order to perform meaningful analysis, work packages (deliverables at the lower level of the WBS) and activities (segments of work) have to be divided (decomposed) into adequate, manageable components. It is also equally important that they have a similar duration. The rule of thumb is that each activity should be approximately eighty hours in duration. Milestones are easier to assess when packages are not longer than eighty man-hours. If an activity is long, an estimation of the percentage of work completed will always be controversial.

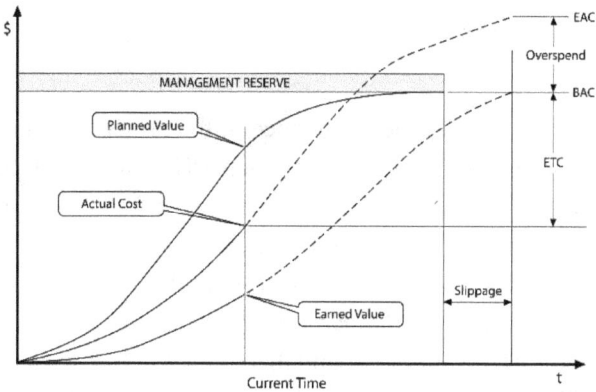

Figure 8 - Earned Value

EAC – Estimate at Completion
BAC – Budget at Completion
ETC – Estimate to Completion

Payments are always linked to the milestones of a project. Milestones are the significant events in the project, and they are usually at the completion of a deliverable. Problems often arise with regards to the definition of completion. There are a few controversial milestones in the construction schedule of superyachts which are linked to payments – one of them is the installation of the engine and drive train arrangement. If the installation of the drive train arrangement is not decomposed into small packages, and if the contract does not clearly state what it means, this could be contradictorily interpreted. I have seen a project where the engine was "installed" and shown to the client. The client paid the instalment and went home happy. A few days later his skipper arrived at the yard and saw the engine had been taken from the boat because a number of other things needed to be done prior to the final installation of the engine.

Interestingly, work packages will be shorter and shorter in the last weeks and days of completion.

The Closing Process

This is the formalizing of the acceptance of the project, or phase, and bringing it to an orderly end. A particularly important aspect of the project's closing process is "lessons learned." In any project there are things which have been done well, but there is also room for improvement. William Bolitho, the journalist born in 1890, wrote in his book *Twelve Against the Gods*: "The most important thing in life is not to capitalize on our successes – any fool can do that. The really important thing is to profit from our mistakes."

One of the areas which are often overlooked is the as-built documentation. It must include all the baseline values measured on the sea trials as well as the drawings, brought up-to-date as "as built" editions, and manufacturer's instruction and maintenance manuals for every single item on board. As well as the technical drawings and manuals, each service needs to have an overview written by the yard's design office, so the crew can get an idea of the general concept of the system before getting involved in the details. Inexperienced yards never take this seriously enough and it is a vital part of a new-build project.

Input – Action – Output

From my observations, not properly following this process (input-output) causes a lot of misunderstanding and leads to wasted time. The gist is that the output of one process forms the input for another process. Computers work perfectly because they follow the input-output process. Those of us of a certain age may remember from the earliest days of the computer era the saying, "Garbage in, garbage out." Input refers to entering information into the system. Output is the information obtained.

This is a fundamental process ensuring the proper flow and use of the information in a project.

This is a basic yet important process that is often completely neglected. The following example explains the input-output concept. For instance:

Example 1 – Estimating process (refer Fig 4)

If expected output is cost estimate for definite estimate (+5/-0%), it is expected that the action required is a bottom-up estimation and to perform this we need an Input: Detailed documentation. By the way, when issuing a cost estimate, always include the "accuracy of estimation".

Stating the accuracy of estimation will prevent the assumption being made that this is the definite estimate or a fixed price. This is particularly important during the early stages of negotiation, when a client is asking for a ballpark figure.

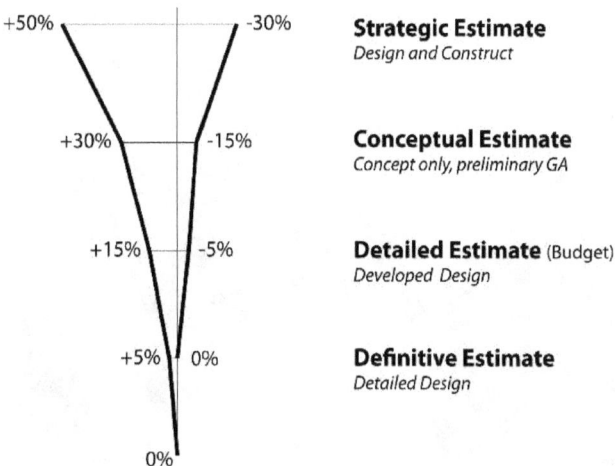

+50% ⋯⋯⋯ -30%	**Strategic Estimate** *Design and Construct*
+30% ──── -15%	**Conceptual Estimate** *Concept only, preliminary GA*
+15% ── -5%	**Detailed Estimate** (Budget) *Developed Design*
+5% ┤ 0%	**Definitive Estimate** *Detailed Design*
0%	

Figure 9 – Funnel Estimating

Example 2 – First an investigation, then a report or business plan

Have you ever seen a business plan done before a feasibility study being conducted? Or, a project charter issued before a business case is completed? There you have it!

STRATEGY THREE

Beware, the half-wise are everywhere

HAVE YOUR HEROES: SCOTT, AMUNDSEN & SHACKLETON

The saying among old Antarctic hands was:

"For scientific discovery give me Scott, for speed and efficiency of travel, give me Amundsen, but when your back is against the wall and there's no hope left, get down on your knees and pray for Shackleton".

Have your heroes too. Have experts you can rely on.

For scientific discovery and technical challenges:

You need a naval architect, marine engineers, designers and knowledgeable suppliers.

For speed and efficiency:

You need a good project manager and a client representative.

When in crisis mode:

You need all of the above, plus a good legal team. Your average attorney won't do the job. First of all, he should be well seasoned in judicial battles. He will need to have a good knowledge of maritime contracts, know the shipbuilding processes and project management methodology, know the Bible inside-out, be able to quote biographies of all famous industrial leaders and know by heart at least three poems of Shakespeare. He needs to have exceptional stamina and plenty of radial energy even when ill.... Well, something like that, anyway.

Something to remember:

Have a mastermind team
A good mastermind team is like a rising tide – it lifts all boats.

"Cathedral-makers, not brick-cutters"

According to an old parable, three men were working hard cutting stone from large blocks of granite. When asked what they were doing, the first fellow said, "I'm making bricks." The second said, "I'm creating a foundation for a large building." The third person answered, "I'm building a cathedral". They were doing the exact same job, and all three responses were accurate, but they reveal each one's difference in attitude. The third person saw the ultimate goal – he accepted the responsibility of the final product. He was able to see the big picture. So, choose your team carefully because a small hole can sink a big ship. Search for top guns. Look for experience, knowledge, attitude and teachability.

10,000 Hours to Competency

Talent seems to play an important role but practice is of more importance. Good piano players practice until they can perform perfectly, but very good piano players practice until they can't make a mistake. Is "the ten thousand hours of experience" a general rule of success? Yes, it seems to be the magic number. Look at Bill Gates, Steve Jobs, the Beatles, and any recognised stage performer or sports player. However, please notice there is a difference between ten years of experience and one year of experience times ten.

Maybe there is truth in the saying, "Old age and treachery will always overcome youth, honesty and skills," as a good friend of mine used to say.

Stakeholders – Beware, the Half-Wise Are Everywhere

"Beware, the half-wise are everywhere." This is a quote from *Hávamál*, a collection of old Viking poems.
One stakeholder may ruin the entire project before the ship is even built, so the management of stakeholders requires a special set of skills of its own. This management requires a lot of what can be referred to as "soft skills." Each stakeholder is different and each one goes in his own direction. Managing stakeholders can be a bit like herding cats. Yachting breeds strong personalities so it should be of no surprise that the stakeholders are sometimes very experienced and always opinionated, but sometimes only half-wise.

Stylist or Optimistic Realist

The stylist (they hate being called "stylist" as in hair stylist, so let's call them designer or yacht designer) is usually the first person approached by a client. A good designer should be not only an artist but also an optimistic realist. He takes the client's requirements, extracts the most important, discards the trivial, and then creates a beautiful and practical solution that is loved by a broader audience. He creates a beautiful yet practical and safe vessel for the purpose. Let's be frank – these days we can see on the Internet or in magazines 3D-created monsters that do not take into account the ferocity and unforgiving nature of the sea. For those dreamers I recommend a book by Czeslaw Marchaj: *Seaworthiness: the Forgotten Factor.*

Naval Architect

Certain caution should be exercised when assessing the ability of someone who calls himself a naval architect. Various people offer different interpretations. It spans from the stylist to the Chartered Professional Engineer (CPEng) or Professional Engineer (PA). By the definition of the Royal Institute of Naval Architects, a naval architect is a professional engineer who is responsible for the design, construction and repair of ships, boats, other marine vessels and offshore structures, both civil and military. The term "naval architect" should not be confused with yacht designer or stylist.

The best yacht designers I know of are those who truly love their work and see it not as a nine-to-five job, but as something they would never wish to retire from.

Here is advice from Paolo Scanu (1954–2008), a naval architect who was responsible for the structural design of s/y *Georgia*:

"Just remember that in a yacht project the best energies spent, both in terms of return on the investment and in terms of fun, are those energies addressed at designing and planning the work before construction begins and remember that a well-designed yacht, attractive, sea kind, fuel efficient, easy to resell, does not cost any more to build than a yacht that, because of being badly designed, is ugly, un-seaworthy, uncomfortable, difficult to maintain, to charter and ultimately to get rid of."

Project Manager (Shipyard's Employee)

The project manager is the individual responsible for managing a project. His aim is to deliver a professional project management service for his principals/employer.

He should have some project management experience and preferably be a qualified (certified) project manager. While representing the shipyard's interests he also must deliver a project which satisfies the client. He walks a difficult tightrope between the demands of the company he works for and the demands of the buyer. This is why some people say that he has two masters or, putting it more bluntly, he lives between a rock and a hard place.

Dealing with a client requires a lot of tact and diplomatic skills at the same time. How vigorously a project manager pursues the yard's rights under the contract is always an important business decision and will always have future consequences, either for him personally or for the shipyard.

If you are planning to be a project manager for a superyacht, be aware – some clients shoot first and ask questions later.

Myths About the Project Manager:

Myth 1 – Project success depends entirely on the project manager.

Even the best project manager can't do the job properly if he does not have a great team at his disposal. Typically a yard is managed by a general manager (GM) and has two to three vessels under construction. Each vessel is managed by a project manager (or project coordinator) who reports directly to the GM.

I would describe the organizational structure of a boatyard

producing superyachts as having a "strong matrix." This means that it is a full-time role and he has a moderate to high level of authority. There are specific functions such as design office, metal fabrication or composite, joinery, fairing and painting, mechanical, electrical, etc. These functions are managed by functional managers. The team is assembled and built up by the GM (perhaps using the popular methodology captured by this slogan: "forming, storming, norming and performing"). So, the skills, knowledge, and experience of the entire team, including project managers, depends on the GM's selection and his leadership. Most importantly, the entire culture of the company is created by the GM.

Myth 2 – Any PM (with IT, civil, government, or agriculture background) can do it.

There is an opinion that a certified project manager can manage any project. He can manage it, but will he gain the respect of the people working with him without very specific boat-building knowledge and not knowing the maritime glossary?

Client Representative

The chief representative on the owner's side is a person nominated by client. This may be the designer, naval architect, broker or any person the client think will best represents his interests. On the surface this title doesn't appear to have a lot of authority on-site, but in fact it carries a lot of responsibility. The client representative is a vital link between the yard and the future owner. The yard's project manager is his prime contact. Make no mistake here: the client representative is not the yard's project manager. Some unexperienced yards allow the client representative to run the project, direct people on the floor and make changes. This is a recipe for disaster simply because there is a strong conflict of interest. His primary roles are:

- To represent the client
- To communicate with yard's project manager
- To make quick decisions on behalf of the buyer if required
- To offer advice on practical, functional or technical matters
- To maintain a record of communication

Shipyard's General Manager

This is the person with whom the client usually meets first when coming to a shipyard and starts building a relationship with the client. His main responsibility is to deliver a good product to the client, but also to look after the wellbeing of the employees and the company affairs. Ricardo Semler, CEO and owner of Semco (Brazil) says it's the general manager's role "to make people look forward to coming to work in the morning." Henry Ford (1863–1947) said this:

"There is one rule for the industrialist and that is: make the best quality of goods possible at the lowest cost possible, paying the highest wages possible."

Eight ways to succeed as a manager:

- Know what your resources are capable of
- Have managerial competence
- Have staff competence
- Be fair and empowering
- Earn the respect of your staff
- Know what you have to know
- Lead and manage (never micromanage!)
- Be accessible and available

Being a general manager is like being the captain of a vessel: "First after God". The character and culture of any business unit depends entirely on the chief executive officer. "Despite the presence of dozens of highly trained specialists on board a submarine, the character of the boat is defined by her commanding officer." —Robert Moore, *A Time to Die: The Untold Story of the Kursk Tragedy*

Note:
Please note, I didn't describe the client's role, nor his responsibilities simply because he, as a client, "has all the rights" and he is above all.

Insurance Agent

If I had to build a superyacht, I would start first from an insurance agent and class surveyor. At the end of the day, or at the day of launching the yacht and raising a flag, the insurance company will have its last say. So, why not ask them at the beginning of project what they need, and how they operate? Having them on board from the very beginning would ensure a smooth ride to completion and help to avoid surprises.

The class surveyor also acts as a guardian of rules and standards and helps from the start guiding the project, thus avoiding unnecessary stress and complication during the design period between the naval architect and the future owner. The classification society surveyor is on the owner's side. He is verifying the engineering quality, thus protecting the owner's interests. In some cases the classification fees are paid by the yard. Personally I would discourage this because it creates a conflict of interest.

Of course, you cannot go to the insurance agent and class surveyor with empty hands; you need some kind of rough description of vessel and couple sketches, so a visit to either a stylist or a naval architect would be beneficial.

Understanding Everyone's Responsibilities

"If a person can take responsibility, he/she can solve almost anything" —Dr Hazel Denning, *Reborn in the West.*

As you can imagine, there are a lot of stakeholders in a superyacht building project, and whether you want them to or not, their responsibilities are often overlapping to a certain extent. To help with the management of stakeholders, the contract should contain an appendix in which the responsibilities of each stakeholder are clearly described. It is up to the parties involved in the contract on how to split the responsibilities. It very much depends on the ability and expertise of each individual stakeholder; however, each function has nominally prescribed certain duties. There are several areas of design causing some confusion. One of these "bones of contention" is usually a lack of clarity around who provides the reticulation plans and one-line schematics.

Often a client expects a complete set of documentation from the naval architect, only to discover that the yard has to further develop this documentation and prepare its own set of drawings. In the opinion of many leading contractors, even the best-prepared documentation from vendors should be converted – and usually is – to the shipyard's drawing system; for instance, the yard's own numbering. Such a conversion provides a good opportunity to review thoroughly each document and put the yard's own stamp on it. This is common practice not only in the shipbuilding industry, and it has nothing to do with violation of copyright. In most of the Western countries the copyright belongs to the naval architect or yacht designer. However, in

New Zealand, copyright belongs to the person who pays for the project, unless agreed otherwise (see New Zealand Copyright Act 1994).

Another area of contention can be the lofting (geometry of vessel in 1:1 scale) and the amount of detail in the drawings. Traditionally, lofting was done manually in the shipyard by an experienced person, but today it can be done on the computer by the naval architect so it then becomes part of the structural documentation. In any case, these issues need to be sorted out before the work commences (preferably before the signing of the contract).

It is very helpful to include in the contract an attachment listing the responsibilities of the main players, whoever they might be.

Something to remember:

You can outsource any job, but you can't outsource responsibilities.

Carrot, Stick and Motivational Speech Won't Take You Far

There are techniques to speed up a project, such as crashing or fast-tracking, but we often tend to take shortcuts. I noticed that sometimes the top brass in project management try to motivate people to work harder. They offer some reward (carrot) and if this doesn't work, they threaten with penalties (stick). This is particularly annoying when a client insists on penalties for late delivery, not realising that the successful completion of a project relies not only on the yard performance, but also on his team's experience, knowledge and co-operation. It also depends on the information he/she supplies, on making quick decisions and limiting the number of changes he requests.

Carrot, stick and motivational speech will only take you so far. Studies show that when the task at hand involves some form of intellectual effort, neither penalty nor reward improves the performance (in fact, they sometimes worsen it). A carrot-and-stick approach helps, however, when physical and repetitive tasks are to be performed.

When the development of a new 30-tonne captive winch for a new 50m LOA sloop is at stake, only brainpower is involved. Certainly for an activity like this, the carrot-and-stick method won't work. However, when the task involves digging a trench (only simple physical power is involved), the stick-and-carrot approach may work.

A motivational speech has a similar effect, unless it is a speech before the battle where it reaches the hearts and plays on the patriotism of warriors. Here is an example of what is probably the worst speech ever given, by a Russian politician speaking to chess players before a chess match:

"You realise the honour that you have to defend. Do you understand the honour? Do you understand properly? Do you understand it or not?" —from the book *Bobby Fisher Goes to War*.

But even if a speech is really inspirational and by some miracle you become motivated by it, the question remains: How would it improve your chess-playing ability or ability to design innovative equipment?

You do not have to motivate nor use carrot-and-stick encouragement for self-starters or clever people with a high job efficiency level. They know what to do, and when and how to do their duties. But if you try to motivate an idiot, he will just do stupid things, only faster. Why? Because, to quote the words of Prince Philip, the Duke of Edinburgh, "The mind cannot absorb what the backside cannot endure".

I would rather follow this philosophy: Positive feedback improves morale; negative feedback improves performance. By the way, in general, people are more motivated by losses than benefits.

Recruitment

I am assuming that for your company you have chosen clever people by selecting them by yourself or through a headhunting process, and not by average recruiters. As the headhunter Nick Corcodilos says, recruiters (with a few exceptions) are trained to recruit recruiters. How can he/she select a person without having specific knowledge? How can they select a procurement manager, designer or sales person, etc., if they know nothing about the job? Relying too much on keywords and mathematical algorithms, they have the potential to discard a lot of great talent.

I personally have a lot of respect for those CEOs who do recruitment by themselves – I mean through his own HR department and his own managers. Your good people are the only thing the competition doesn't have. Why would you entrust such a vital process to someone else and not do the job yourself? Besides, recruiting agencies constantly complain that they have so many applications, they can only devote six seconds to each CV, etc. They work for many companies and they constantly advertise for a number of positions, while a human resource manager on-site may have only a few positions to fill so they can afford to spend more time reading CVs. They can answer queries and they can afford to have several interviews with a potential employee.

So who are the star performers? They are strong leaders who get intended results regardless of the challenges. They're highly motivated to do the actual work required. They take on projects no one else wants and they fit seamlessly with the people, culture and manager.

Here are a few practical and tangible points to look at when selecting a project team (on shipyard and client side):

- Are they well qualified?
- Are they experienced?
- Have they done a similar boat before?
- How well they know the local and international market for supply of equipment?
- How big is their professional network?
- Do they have enough expertise to make fast decisions?
- How will they fit into the existing team and do they have a positive attitude?
- Listen to your subconscious.

Tomek M. Glowacki

STRATEGY FOUR

"Succeed together or fail apart." —Leslie Gelb, *Power Rules*

BEFORE YOU GO TO THE ALTAR

The client and the yard need to assess each other carefully before making a selection, the yard selecting the right project and the client selecting a suitable yard. This is because, once the contract is signed, it is like a marriage: you have to steer the course together towards a mutually chosen destination.

To help you with this process, below are two sections advising the client and the shipyard on how to choose each other:

Advice to the Shipyard: Choose the Right Project

There is a saying amongst project managers: "Choose the right project". Sometimes it is easier said than done. Not all yards have the freedom of choice due to the fact that they do not have a lot of clients knocking on their door. In fact, today, yards are desperate to get any deal. There is no longer any choice involved. They take what they can get! Whilst a yard is in control of, and can therefore improve over time its quality and performance, such things as the exchange rate and the global economy are out of its control.

Here are a few practical and tangible points to look at when selecting a project:

- How the negotiation process goes. Can you see anything weird?
- Is the project well defined? How many drawings and of what quality does the client provide with the contract?
- What support will we get from the external vendors such as the designer, naval architect, and structural engineer? Are they experienced? Have they done a similar boat before?
- What profit can we expect from this project? If it's anything less than 20 percent, alarm bells should ring.
- Do we have the resources and expertise required for the successful completion of this project? This is particularly important when moving to construction of a much bigger vessel than anything previous.
- Does the client or his team have enough expertise to make fast decisions? The client may not be too experienced, but in this case he has to have an experienced skipper, chief mechanical/electrical officer, etc.
- Does the client have sufficient funding? Perform a credit check!
- Listen to your subconscious – your gut feel. There may be something telling you, this project is not OK.

For instance, one of my potential clients revealed a bit of his past in the initial stage of negotiation. He said, "In my bar at home I have seven stools. Each one comes from a different pub. In order to steal the seventh stool, friends of mine had to help me. I had to pretend to be a person with a paraplegic disorder and they did the rest of the job". Loud alarm bells began to sound.

Listen to this story this is a gem:

A friend of mine (a yacht designer) met his potential client in a marina yacht club to discuss a potential project. After a short discussion, the potential client invited my friend to his boat which was moored just a few hundred yards away in the marina to show him certain features which he wanted to replicate on his new project. They took the dinghy and paddled to the yacht. They decided to go for a short afternoon sail. After starting the engine they realised the steering wheel was missing. They looked around and on the balcony of the yacht club there were three men with his steering wheel. "Pay us our money and you will get the wheel".

Advice to the Client: Choose the Right Shipyard

Choosing the right shipyard may seem like an easy task, but as history has shown us, it is not. An incorrect decision can result in a calamity. So, how can we assess the capability of the yard and predict its performance? For starters, talk to the shipyard's existing clients if you can obtain access to them. If they are willing to talk to you, ask them questions such as: What was the percentage value of variation orders (VOs) to the total value of the boat? How long did the yard take to do sea trials and commission the boat? In other words, how long did it take from launching to handing over? What is the general attitude of the people involved? Do they have a "can do" attitude or is there blaming, excuses and a culture of denial? What level of technical expertise do they represent? Is there a strong *esprit de corps* – a sense of unity and common interests and responsibilities? How good is as-built documentation and vessel operational instructions? And, most of all, how big and how experienced is the on-site design office? If the shipyard's existing clients had a bad experience, they may not want to talk on this subject and wish to forget about their ordeal altogether.

Look for a relatively smaller shipyard. Jack Welch, former CEO of General Electric, one of the biggest corporations in the world, used to say, "Behave like a small company". He understands the value of being small: communication is less formal, sales talks directly to marketing, marketing talks directly to design, etc. There is less paperwork; a lot of ideas are discussed at the watering hole and the decision-making process is faster.

How small is small enough? For some companies it is five hundred employees, but for others it could be one hundred. It depends on the type of industry. The thing is that people will perform better (at their potential) if they know almost everyone around them. For some reason one hundred and fifty employees seems to be the magic number. Gore Industries (they make the famous Gortex) is made up of several companies. When one company grew above one hundred and fifty people, they split into two small units and built a new facility with a car park for only one hundred and fifty cars. Ricardo Semler came to the same conclusion; Semco (Brazil) tends to keep production units to around one hundred and fifty employees.

I have worked in boatyards with as few as thirteen people and in shipyard with as many as thirteen thousand people (behind one fence) and everything in between. There is no doubt in my mind that smaller units operate more efficiently; people are better utilised, leadership is easier and the quality management system can be simple.

Another way to get a quick feel about the yard is using methodologies such as "the five S's" or "Rapid Plant Assessment". This is nothing more than walking through the yard, observing things carefully and filling out the set of questions.

STRATEGY FIVE

Clear requirements = excellent product

"IF HE'D LET ME KNOW…"

Project Definition

I cannot emphasize enough how important it is to get the definition of any product (or service) right before commencing the project.

"If he had let me know what he was going to do with the boat, I would have built him the boat to do it". —Rough Passage

This is response of the designer of s/y *Emanuel* Andrew Anderson's to some of the complaints of Commander Robert Graham (writing to him from Newfoundland) on *Emanuel* after crossing the Atlantic. *Emanuel* was the successor of the yawl s/y *Caplin* and was built for an around-the-world voyage but not for Newfoundland waters – an area of icebergs and hard weather. Of course it could be debatable as to whether a boat suitable for an around-the-world voyage should not easily withstand sailing around Newfoundland, but the point I am trying to make here is that the purpose and function of each product has to be very well defined.

Proceeding with a project without a well-defined scope of work and technical specifications has regrettable consequences. Here is an excerpt from an article by Ivy F. Hooks, president and CEO of Compliance Automation Inc., regarding the relationship between the definition of a project and cost overruns.

"People who write bad requirements should not be surprised when they get a bad product – but they always are. Almost everyone is aware of problems encountered by the NASA Space Station Program. The Space Program made a mistake that no project can afford to make – it did not define the scope of the project before the requirements were written. NASA data shows that cost overruns of 100% to 200% are common for those projects that spend 5% or less of project cost on the front end. Projects that spend 5% to 10% up-front show maximum overruns of 120%. Those that have invested 10% or more show zero to 50% overruns. Failing to invest up-front is penny wise and pound-foolish. In direct contrast, Boeing Aircraft has been very successful with their 700-series of aircraft. They have typically spent 15% on the front-end effort, defining the scope, developing the requirements, and doing the preliminary design. They believe that they can invest more and save more and to that end have invested 30% up-front on the 777 aircraft. Those who try to short cut this front end effort have this motto: "We don't have time to do it right but always have time to do it over". If you are experiencing a large number of requirement changes during development, you need to look at how you have defined the scope. It is very likely that you have a poorly defined scope and that many bad assumptions have been made in writing the requirements. The process begins with defining a need. The goals, objectives, constrain, operational concepts, and a high-level of definition of the system is the output of the process. Other items that need to be defined are budgets, schedules, and management."

Some other comments confirming the above:

"On capital project the single best payoff in terms of project success comes from having good project definition early." — Rand Corporation

"As we all know, each client is unique. Each is in the game for a slightly different reason, and each is trying to obtain something special from his project. As professionals, it is imperative that we take the time to find out more than just what he wants in his boat. We must find out what motivates him, if for no other reason than to be in a much better position to avoid alienating him as the job progresses". —Mark T. Masciarotte of DSG Association

Something to Remember

There is only one conclusion here: define, define, and define the product and scope of the project at the earliest beginning.

Getting to Know Each Other and Closing the Deal Fast

I'm aware of one project where negotiations and the starting and stopping of a project lasted for fifteen years. Such an extreme case doesn't do any good for cash flow, self-esteem or the morale of all parties involved.

Getting to know each other can be relatively easy, but closing the deal fast can be tricky. If neither money nor technical issues come into play, the question is, how fast? When do we close the deal and start the project?

The general opinion is that once the information is in the range of 50 to 70 percent, you can start the project. But how can we recognize that 50 to 70 percent of the information is on hand in the case of a superyacht? And, is 70 percent of the information enough? Here is my definition, which isn't tied to any percentage in a particular case. I would start a project when:

- All geometry is defined (hull and superstructure).
- General arrangement is completed (interior layout).
- Deck layout is completed.
- Construction drawings (structural) are completed and approved.
- Weight study completed 95 percent.
- Hydrostatics and stability calculations are satisfactory.
- Space allowances are checked and confirmed.
- Main equipment is selected.
- Technical specification is completed 99 percent.
- Tank testing completed (optional).
- Classification drawings are approved by Classification Society.

The above information can be presented by 20–25 drawings and described in 30–50 pages of a specification book. In my experience, this is adequate enough information to begin construction; however, it may differ from yard to yard depending on in-house skills and expertise.

With superyachts, product development starts with design brief well before signing a contract with a shipyard and continues until launching, although it will slowly fade away.

Something to remember:

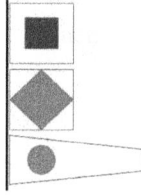

Figure 9 - SF1: Are You Ready to Get Underway?

If You Do Not Know How to Ask the Right Questions, You Discover Nothing

"You should not ask questions without knowledge and if you do not know how to ask the right question, you discover nothing".
—W. Edwards Deming (1900-1993)

Request for Proposal (RFP), Request for Quotation (RFQ) and Request for Information (RFI) are common forms for eliciting proposals, quotations or information from potential vendors for products or services. Each one has the potential to be eventually converted into a purchase order. Because the procurement of materials, equipment and services are a big part of the project management process, the quality of these documents is very important. I wouldn't write about it had I not seen terrible examples of these types of documents. In one case the description of the product was as follows: "Gearbox 1 ea". Of course, there was an earlier discussion with the supplier as to what type of gearbox, when it should be supplied, etc., but the problem is that this document does not leave any trace of information for the future if anything should go wrong.

In order to receive the right answer, the right question has to be asked and the right information has to be provided. Each requisition has two parts: technical and commercial.

Technical: Prepared by project manager/design office

Section 1 – Scope of Supply: work included, work excluded and battery limits (boundaries of supply)
Section 2 – Technical Specifications
Section 3 – Drawing and Data Requirements
Section 4 – Special Project Conditions
Section 5 – Quality System Requirements

Commercial: prepared by purchasing and legal department

Section 6 – Expediting
Section 7 – General Conditions
Section 8 – Shipping and Packaging Instructions
Section 9 – Invoicing Instructions
Section 10 – Form of Proposal

It is not good enough to pull out a couple of pages related to a mast or electronics from the general specifications and send them to a contractor asking for a price. Disapproval guaranteed.

STRATEGY SIX

Zero in on the requirements

THE DESIGN SPIRAL

Yacht design is an iterative process, similar to aircraft, tank, and software development. It relies on a "trial and error" approach. A designer starts with a number of assumptions and works through the design process, changing these assumptions until he achieves the desired result. Each cycle involves more calculations and detailing, thus increasing complexity of the design process but at the same time reducing possible design choices. It is a general concept that the yacht designers modify to suit their own design sequence, helping them to resolve the major trade-offs. The sooner the designer resolves all trade-offs (compromises), the sooner he is able to complete the final drawings for construction. The design process of any vessel is long and elaborate. The better the design brief, the better the definition of product, the more detailed requirements – the fewer number of circles.

Design Process & Documentation

Step 1   Design Brief
Step 2   Conceptual Design Phase
Step 3   Preliminary Design Phase
Step 4   Technical Designs
Step 5   Documentation Approved for Construction
Step 6   As-Build Documentation

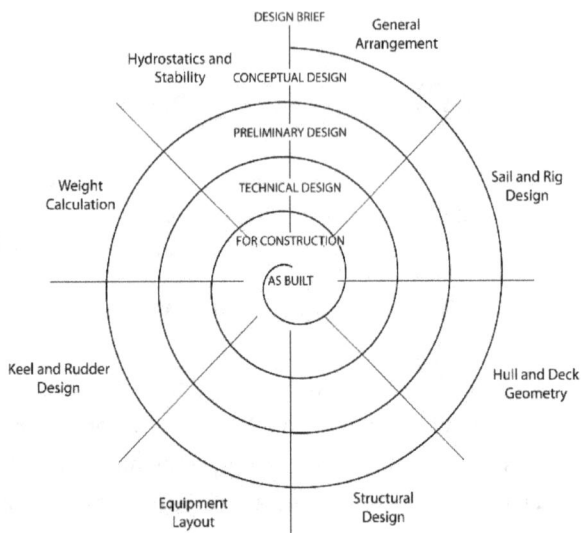

Figure 10 - Design Spiral

Something to remember:

Luck is the residue of good design
- Branch Rickey (1881- 1965)

The Design Brief

The design brief clarifies the purpose and goals of the vessel and consists of the following parts:

- Scope statement, or mission of the vessel
- The owner's design requirements
- The design constraints

This should be written, reviewed and signed off on by the owner before the concept design stage is begun. This sounds a bit formal but it ensures smooth co-operation between a client and a designer. It is in the designer's interest to stimulate and help to prepare such a brief.

Scope Statement or Mission of the Vessel

The first step in the design of any product is to define very clearly its main function or purpose. A boat is no exception. This clear idea of how the boat will be used is what we call a design statement. It defines the main function of the boat and is used as a vision to guide you through the various trade-offs that must be made to achieve the final result.

The scope statement (although often ignored) plays a very important role. It defines the purpose of the vessel using one sentence or paragraph. It helps the designer and client in making future decisions, to stay on track. It helps when making decisions on trade-offs (compromises) during the design and construction of the vessel.

Samples of a design statement:

"A home on the water for my children".

"60-foot sloop or cutter for a retired couple to live on board all year round".

"50-foot sloop for cruising and racing: pace, grace, and space".

"A small ferry to carry passengers between Hobsonville and Davenport, in a safe, fast and comfortable manner, that will maximize profits over the life of the vessel.".

"The objective is to accomplish a solo, non-stop, easterly circumnavigation by sail in 180 to 220 days. The boat design challenges are implicit: A boat to bear the rigors of six months'

continuous sailing in varied weather conditions, much of it heavy; a boat to average 6.25 knots or better over that span of time; a boat that will match the age and physical conditions of the skipper; a boat that will accept the required stores and equipment and kindly give up those stores one day at a time; a boat that has as few structural and mechanical weak lines as possible; a boat that is planned with redundancy in key systems and equipment; a boat that uses mechanical and electrical advantages, proven engineering solutions, yet can be sailed without them." —Dodge Morgan, s/y *American Promise*

As you can see, a design statement can be short or very elaborate. Please also note the difference between a design statement and a project statement. A design statement describes a product, while a project statement describes a project, including scope, duration and budget.

The Owner's Design Requirements

The owner's design requirements usually consist of the following parts:

- A list of design requirements. Here, the client may list all major attributes of the vessel and their ranges in decreasing order of importance. For example: number of cabins, speed and cruising range.
- A checklist of design options with weight factor (a "classification"). For instance: "must have" – factor 3, "desirable" – factor 2, "would be nice to have" – factor 1.
- An owner's description of how the boat will be used. If the owner will tell you how he is going to use the boat, you will be able to design him a boat to do it. Perhaps this is where the whole design should start. This way, the designer would

get to know what the client's needs are, giving the designer full freedom in choosing the best design option.

- Photos and descriptions of other boat features that the client would like to see on his boat. This is self-explanatory: a picture is worth a thousand words. A complex idea can be conveyed with just one picture.

The Design Constraints

The design constraints include any limits over which the designer has no control. It includes constraints that are imposed on the design by the expecting environment, such as height limit for clearance under the bridges, draft limit for shallow water, and beam for canals, or by outside organizations.

The Conceptual Design Phase

The conceptual design or design proposal is the phase which could be contained in a couple of drawings showing the general arrangement, sails and rigging, and includes a short specification giving general dimensions. This is kind of a feasibility study aiming to show the client that his requirements can be fulfilled. If the client accepts this general concept, the designer, after signing a contract, moves to the next design phase, the preliminary phase.

The Preliminary Design Phase

In this phase, the designer determines more details and performs more calculations. In fact (although all these phases are arbitrary), this is where the whole boat emerges: the hull shape, deck geometry, interior layout, equipment layout, scantling, sail and rigging plan. Calculations are performed, including hydrostatics, stability and weight studies. The designer may be required to repeat this process several times to make sure that all the parameters correspond with each other. At the end of this phase documentation is usually passed for approval by the classification society and sent to several shipyards with a Request for Quotation (RFQ).

The Technical Design Phase

The classification society and shipyards may have some comments, suggestions for improvement or simplification, or demand changes to satisfy the rules. When the documentation comes back from those reviews, the designer now implements the recommended changes and produces the final deliverable to the client. This documentation can now be included in the construction contract. However, this is not yet the documentation "for construction".

"Approved for Construction"

At this point the documentation has changed hands from the designer/naval architect to the shipyard's design office. Now it is the responsibility of the shipyard to review and complete the documentation and produce detailed designs – "shop drawings" – which can be issued with a stamp: "for construction". Only drawings with this stamp have the right to be on the shop floor. Trust me, this small red stamp makes a big difference.

The obsolete drawings in production cause a lot of confusion. Therefore, when drawings approved for construction are revised, all previous drawings issued to the floor need to be collected and destroyed, but still keeping all versions in the design office.

Check, Review, Verify, Scrutinise, Approve and Check Again

The good definition of a project and attention to detail are the keys to project success. Attention to detail can be achieved by relentless checking, reviewing, verifying, scrutinising, approval and checking all over again.

In 2009 the Institute of Professional Engineers New Zealand (IPENZ) and the Association of Consulting Engineers New Zealand (ACENZ) jointly issued a practice note on "Structural Design Office Practice". This note is applicable to any type of engineering work and includes guidance such as:

- Follow logical design phases – concept, preliminary, developed, detailed – with review ... at the end of each phase.
- Involve senior and experienced engineers in deciding the structural form.
- Some form of quality assurance or internal review process is essential to ensure consistent and defect-minimised design output.
- For large projects, frequent design review, and designated hold-points are essential.
- While arithmetic accuracy can usually be checked by competent junior staff, other aspects of the review invariably require input by experienced professionals.
- Small practices and sole practitioners need to be particularly mindful of how to achieve effective review, particularly when undertaking complex work. Some form of external review, possibly on a reciprocal arrangement, may be an appropriate solution.

- Effective, detailed and thorough review of drawings is a tedious yet essential task, usually requiring input by senior staff.

Something to remember:

The designing of a superyacht is a long, iterative and elaborate process and it doesn't tolerate any shortcuts.

STRATEGY SEVEN

Always measure your progress

YOU CAN'T CONTROL WHAT YOU CAN'T MEASURE

Unless you measure changes, you don't know if the project is getting better or worse. In yacht design, due to the fact that we constantly deal with trade-offs of one feature for another, measurement of values representing that feature is critical.

You've probably heard this before: "We design and build yachts without compromises". What nonsense! There is no such thing as a yacht without compromises. The construction of a yacht (also aircraft, a tank, a car, etc.) is one big compromise. Please let me explain:

We all know that weight and waterline length are defining factors for a boat's speed. If the client wants to load the boat with added conveniences, the weight will go up. To maintain the same speed as before the change, a bigger engine is needed. To maintain the same range, more fuel will be required, possibly heavier ballast, perhaps the waterline should also be extended, and the trade-offs can go on forever.

- If you are adding more ballast for greater stability, you make a boat heavier and therefore slower, so you need more sail area – you compromise.

- If you make the waterline surface fuller and therefore the moment of inertia bigger, you increase stability but reduce speed – you compromise.
- If you would like to install a bigger engine for more speed or torque in rough weather, you may add some fuel for the given range, but you reduce speed and space inside – you compromise.
- If you decide to have a lifting keel, you may have a low draft but you take up space inside and add to maintenance – you compromise.

To achieve the desired outcome, we follow an iterative design spiral process. During this process sometimes we are closer to, or sometimes further from, our goal. This is why we need a method of measuring our progress. Some compromises outweigh others, so to keep these gains and losses under control (measurable), naval architects (and also aircraft designers, tank designers and, in a big way, military planners) use a methodology called "Measure of Merit".

Measure of Merit – MOM

Among engineers there is a perception that, before you start calculating something, it is good to estimate an outcome of this calculation. This has a dual purpose. First, after a while we learn to estimate without calculation. This is a priceless skill, saving a lot of time. Second, if the result of the calculation differs a lot from our estimate, it is a warning that something is not right and it forces us to check again. As wisely stated by Winnie the Pooh, "It's best to know what you are looking for, before you look for it."

A Measure of Merit for a vessel is a specific formula that

converts the complete design into one number that tells you if boat design "A" is better than boat design "B", and helps you select between major trade-offs.

This helps to identify many of the issues associated with the selection, development, application and interpretations of measures, criteria and standards, and to help solve the problem of generating these measures. Its particular purpose is:

- Assessment and evaluation
- Integration with other elements and attributes
- Prediction and forecasting
- Discovery
- Recordkeeping
- Regulatory

It answers the following questions:

- Is it good enough or better than what we already have?
- Is it getting better or worse?
- Is it within acceptable limits?

Superyachts are one-offs. They are designed and built to the specific requirements of the client and the client wants to know whether this next version of its design is better than the last one. This is why designers translate the purpose, or mission of the vessel, into a mathematical equation that helps you to compromise one feature for another. Measure of Merit is possible for all kinds of crafts: commercial, racing yachts or pleasure boats. This method is also widely used by militaries to choose, assess and support strategic decisions.

Note:

It is beyond the scope of this book to explain how to calculate MOM for each type of vessel and how to optimise design. You have to make an individual approach and customise calculation for a specific vessel, client or circumstance.

STRATEGY EIGHT

Keep project constraints relevant

WE DO THREE TYPES OF WORK HERE

Project constraints are any limiting factors that affect the project's execution or process. Don't confuse project constraints (scope, duration and cost) with design constrains such as draft of the vessel, height of the mast, etc. Traditionally, the project management community recognised only three project constraints: scope, duration and cost. However, recently another two constraints have been added: quality and safety – and rightly so. Projects without quality make no sense, and profit without safety has no value.

Some projects have to be completed on time at all costs; for others, quality is the first priority and for some, it's the cost. Not all project constraints have the same weight at the same time.

Time, Cost or Quality - Driven Project

When you ask a client what is more important to him, the date of completion of the project, its quality or the money spent on the project, he will usually say all of them are important. Tell him this very old theme: "We do three types of work here: good, fast and cheap; however, you can only have any combination of two". I will try to explain it using the following three scenarios:

Tomek M. Glowacki

Scenario 1: Funeral of Turkish president Turgut Özal

Time is of the utmost importance!

It is April 17, 1993, Istanbul. With pomp and ceremony fit for a sultan, Turkey is laying its reformist president, Turgut Özal, in his final resting place. Organizers are expecting tens of thousands of people to attend the state burial ceremony, in which Özal is going to be buried next to the mausoleum of Adnan Menderes. Tens of thousands of Turks are expected to march for two hours behind Özal's coffin from the mosque of Mehmet the Conqueror to his green granite grave outside the city's Byzantine walls. Organizers have a mammoth job ahead of them to prepare the site: roads, car parks, sanitation, stairs, flowers, plants, lawns, etc. All have to be ready in ninety-six hours. The budget for this project is $3.5 million, but nobody will argue if costs overrun. Cost is not so important in this project. Also, it is acceptable that the quality is not perfect. Everybody knows that after the funeral some additional work will have to be done, but readiness of the site is obviously the first priority in this case.

Here is an example from the boating industry:

The completion of any racing boat for a race where the date and the hour of the start are set. In this case, being one hour late is not an option. I have seen yachts leaving the marina with a pile of stuff on the deck and a guy on the mast setting the rig. It is obvious that the boat is not completely finished and the quality is debatable. Nobody at this stage is asking about cost, either. The yacht is going to race in the national championships. Being on the start line is the most important thing. Time is of the essence. Nota bene, in this case she won the 1st IOR division, so the quality was acceptable.

Scenario 2: "We choose to go to the moon" —JFK

In this case quality (as you can expect) is of the utmost importance.

On May 25, 1961, John Fitzgerald Kennedy proclaimed, "I believe that this nation should commit itself to achieving the goal, before this decade is out, of landing a man on the moon and returning him safely to the earth."

Launched by a Saturn rocket from Kennedy Space Center in Merritt Island, Florida, on July 16, 1969 the *Apollo 11* astronauts – Neil Armstrong, Michael Collins, and Edwin "Buzz" Aldrin Jr. – went to the moon. On July 20, 1969, they returned to Earth and landed safely in the Pacific Ocean on July 24. They had realized President Kennedy's dream.

In this case, obviously quality was the most important thing. Nothing dramatic would have happened if the launching of the rocket was postponed by a couple of weeks or months or even years. Nobody would argue if the final cost was blown out by a few million dollars. In this case, getting the astronauts on the moon and returning them safely was the most important objective.

An example from the boating industry is the completion of a superyacht. The quality is the most important thing, to the degree that a malfunctioning coffee machine can ruin the owner's holiday (true story – take my word for it). A superyacht's cost can be, and is almost always, a bit higher than estimated. The schedule can overrun, but what is the point of having a superyacht if the equipment is malfunctioning?

Scenario 3: Family holiday

A family holiday can be a perfect example of where the most important constraint is cost. We can have a lot of fun and there are a lot of possibilities to satisfy our demands. Accommodation could be a three-star or four-star hotel, backpackers', homestay or couch. Then there is food and entertainment. But money can be, and in many cases is, restricted.

An example from the boating industry: Any commercial vessel, if she cannot deliver the required return on investment, is not worth building.

Something to remember:

Quality, cost and schedule rarely have equal weight of importance.

STRATEGY NINE

KISS – Keep it Short & Simple

"WRITE TO BE UNDERSTOOD, SPEAK TO BE HEARD"

"Write to be understood, speak to be heard, read to grow." — Lawrence Clark Powell

Mark Twain once apologised for a long letter, because (as he explained) he didn't have time for a short one. Indeed, if you can't explain on one page, you don't really know what you want to say. Less is more. These days many companies aim to write only short memos and letters. The "headline memo" is becoming a standard in modern communication. The aim is to contain the whole message in the title. The rest of the page is just a short intro, body and conclusion. A shorter letter, getting right to the point, has a better chance of not losing your audience. Here is a hint: according to industry executives, an important, well-structured letter on an executive level usually takes the whole day to prepare and is contained in one A4 page.

Each time you take a pen to write or open your mouth to speak, you are trying to get into the mind of your reader or listener. He is tuned only to a very selective frequency. If the message you broadcast is not on that frequency, the message will fall between the desks into a "round filing cabinet" – the rubbish bin. The frequency for your reader or listener has to be WII-FM (What's In It – For Me). Keeping a discussion on subject, particularly during a project meeting, is a task in itself. Have you ever been involved in, or at least followed, a discussion on the Internet? There you have it!

"If you can't explain it simply, you don't understand it well enough." —Albert Einstein

Something to remember:

All communications regarding superyacht construction
ultimately have to go through two people:
The project manager and the client representative

Documents under Control

While it is one thing to write and receive correspondence, it is another to keep good track of it. This is particularly important in the era of emails – see the next chapter. In order to regulate the flow of important information (distribution, approvals, and revisions), I strongly advise on the creation of a "documents under control" procedure. The "documents under control" procedure is designed to ensure that:

- All documents are appropriately actioned, reviewed and approved prior to issue,
- The current revision of documents are located where they are readily accessible to document users,
- Obsolete copies of documents are removed from use,
- Documents are distributed and filed in a controlled manner, and
- The production of forms is controlled.

Read it, understand it, act on it, delegate it, file it or throw it away. Don't be afraid to throw things away, but contemplate first, like Alfred Sloane CEO of General Motors did, "What is the worst thing that can happen if I throw this out?"

A small detail, like the method of filing, can make a big difference – what document to file, how it should be filed, how documents should be linked together, indexing, etc. Once again, small things can make or break a project. One of the most respected long-distance sailors, Bernard Moitessier (1925-1994), believed that "a wrongly placed matchbox can cause a disaster." It's all about attention to detail.

Emails vs. Letters, Memos and Forms

Emails differ from letters, memos, business forms and templates. They are more casual, and they often carry more than one piece of information, question or request. Emails do not have "prompts"; therefore, they do not force or at least guide a respondent to answer in a structured, formal manner.

If communication is not formalized, it leads to disarray. As an example, let's take a document called a Technical Query (TQ). A TQ starts with a template. It is prepared once and used over and over. The requirements for the originator and respondent are laid out, forcing them to fulfil all queries properly and give an adequate answer. First of all, such a template is assigned a document number, which allows it to be properly categorized and filed. Second, it forces the writer to satisfy all prompts. Third, it prompts the respondent to give the required information. In contrary, the tracking of information contained in emails is particularly difficult, if handled by several people.

Some people like emails and some don't, each group for different reasons. It's a controversial subject, but an important one. In some companies emails are the primary form of interaction and they become counterproductive. When discussions start involving too many people, misunderstanding becomes amplified, it is not clear who should make a final decision and things are put on hold. Have you also noticed that emails can sometimes create an adversarial environment?

I would say emails are great, if used on the right occasion and with a bit of attention to detail. Email is only one of the many tools at your disposal. Too often it replaces a real conversation or a formal document. Let's start with the hierarchy of communication in reverse order:

- Discussions at the "water cooler"
- Phone conversations
- Meetings
- Emails
- Letters and memorandums
- Formal documents created on templates
- Reports
- Technical documents, drawings
- Contracts, agreements, etc.

All discussions may be legally binding, provided that they are confirmed in writing. Emails become more and more in use, and in a country like New Zealand, they become a legal document. Therefore, the content should be treated very seriously. There is nothing wrong with using email if it is used properly, you adhere to email etiquette and pay full attention to detail. But unfortunately, this is not usually the case. As I have observed, emails are used as a conversation tool, for loose chatting. Even if they ultimately lead to solving some project problems, the sheer amount of exchanged emails and number of people involved make it difficult to follow and reference. In communications like that, someone needs to sort the wheat from the chaff. It is critical to follow up closely with what is actually being said and agreed upon.

A project manager's duties can be harder to manage when business is being conducted through emails. Such discussions should eventually end up in a formal document such as specification, contract, status report, schedule or decision log, change request etc. If this happens, that is OK, but if not the actual message may not be communicated clearly. Some people believe that emails provide an audit trail and a documented reminder of what was agreed to. This is true, but if a discussion is not properly concluded in a project management template, following the trail is time-consuming and very difficult, if not impossible.

Understand Documents and Their Legal Hierarchy

Not every document has the same weight of importance. The question is, which document should take priority? Various industries have differing opinions or have no opinion at all. In general in the superyacht industry, the hierarchy of documents is as follows: contract, specification, contract drawings, and correspondence. However, the best way to know how one document will take precedence over another is if it is spelled out in the contract.

Common "Deadly Sins" in Contract Writing

From my observation these are the most common areas neglected in the preparation of contract documents:

- Lack of table of contents and definitions.
- Technical jargon, legal lingo, ambiguities and aggressiveness.
- Long sentences
- Lacking in grouping by subject.
- Terms and conditions of contract, scope of work and specifications are mixed together.
- More than two parties to the contract (unless it is an Alliance contract).
- Agreement favors one party (take-it-or-leave-it) and doesn't foster a win-win scenario.
- Wordy and lacking in clarity.

Something to remember:

A contract is not only "the lawyers' talk".
You, as the signatory, have to understand every word in it.

The Contract and Its Structure

Legal lingo is only part of the problem. What I often notice is that, habitually, all the different parts of a contract document are mixed together. This makes the preparation and the execution of a contract very difficult. In contrast, when all the various parts of a contract agreement carry only the information relevant to that part, each part can be prepared separately by the relevant person – for example, legal by the lawyer, specifications by the engineer, scope of work by the project or contract manager, financial by the accountant and so on. Of course, in the end, all the documents have to be reconciled.

There are many types of contracts and they usually have a risk-sharing formula between the client and the contractor. This in turn is reflected in the profit, so choosing the right contract can make or break a project. This is how Butch Dalrymple–Smith of Butch Design recall his experiences:

"Build contracts can take various forms:

- *The very large shipyards tend to ask a fixed price. In this case there is no incentive for the owner or the owner's representative to be reasonable. He will reject any workmanship that is not 100% excellent (even if it is invisible) and will insist on meeting the exact letter of the specification even if some things are better and some slightly worse than the quality specified.*

- *Small yards with little experience of pricing yachts, may ask for a "time and materials" contract, where the yard simply says that every man-hour will cost $x and will mark up all materials (& subcontractors) by 10%-15%. In this case there is no incentive for the yard to be reasonable. In theory they can take longer to do the work and get paid more for it. There is no incentive for them to be either economical or efficient.*

- *A compromise is a fixed profit contract (as used to be practiced at Royal Huisman Shipyard). Here the yard charges their man-hours and materials at cost and then adds another item for profit (as they are professionals and they do not want to do the work for nothing). Of course the definition of a man-hour "at cost" can be variable (does it include the cost of the premises, are the secretaries costs added in, etc.?) However, I have seen Wolter Huisman negotiate on this basis and although the clients used to argue about everything else, I have never heard any of them query the amount the yard owner believes he should profit from his activity. Another form of this contract (Camper & Nicholsons, when I was there in 1997) is that we would estimate the number of hours necessary to build the yacht and sell those hours at a price that gave us a good margin. Any hours spent over this ceiling would be charged at cost. This is exactly the same, but wrapped in a more genteel package. Note that in both cases there is an incentive for the yard to complete the yacht in the lowest number of man-hours and yet neither side is penalised if the specification is changed during the build."*

The major components of a shipbuilding contract are:

- Agreement
- Scope of work (deliverables)
- Terms and conditions
- Technical specification
- Contract drawings
- Regulations
- Classification rules
- Standards
- Execution page

Note:

In New Zealand, even a verbal agreement should be honored, but the problem is that it is difficult to defend. Many people in the business agree with film producer Samuel Goldwyn, who says, "A verbal contract isn't worth the paper it's written on."

Something to remember:

A contract is about sharing risk between the client and the
contractor. Weigh every word in it.

Specification

Technical requirement for superyachts are contained in the
technical specification. There are three types of specifications, or
a mix of them:

- Design or end product specification
- Performance specification and
- Process (or procedural) specification

Figure 11 - s/y *Quintessential,* 100 ft LOA cat,
Warwick Yacht Design Ltd

The requirements of each carry different obligations. A superyacht's specifications are always a mixture of design and performance specifications, depending on what part of the vessel it describes. For instance, the type of specifications for fairing and painting will be different from the type of specifications used for a drive-train arrangement.

Problems may arise when, for one part of the vessel, two types of specifications are used. If this is the case, there is a danger of complying with one spec but not complying with the other. For instance, the client may specify particular equipment (e.g. main engine and propeller) and at the same time, specify a particular level of performance (e.g. speed of vessel). It may happen that this engine cannot deliver the required performance for this particular hull form.

I would strongly advise, in specifications for superyachts, to use as much as possible the design/product specifications type.

Something to remember:

In specification, the room for interpretation should be as little as possible.

Sample of mixed specification:

Genesis 6

"...So make yourself an arc of cypress wood; make rooms in it and coat it with pitch inside and out. This is how you are to build it: The ark is to be 300 cubits long, 50 cubits wide and 30m and 30 cubits high (450x75x45 feet). Make a roof for it and finish the

ark to within one cubit (18 inches) of the top. Put a door in the side of the ark and make lower, middle and upper decks. ... You are to bring into the ark two of all living creatures, male and female, to keep them alive with you."

The boat was to be built to specific dimensions, (design spec), but she will need to accommodate all living creatures, (performance spec). What would happen if the design spec did not complement the performance specification? In this case we were lucky as the engineer who prepared the specification was exceptionally good.

Something to remember:

To avoid chaos, communicate with clarity.

Tomek M. Glowacki

STRATEGY TEN

Treat the causes, not the symptoms

KEEP PROFIT HIGH AND COST OF QUALITY LOW

Good profit (financial benefit) is the result of good performance. Technically speaking, it is the difference between revenue gained from a business activity and the business expenses needed to sustain the activity. Any profit gained is looked after by a board of directors on behalf of the business's owners. They decide how to spend this money. For example, do they put it back into the business, or invest in another business? By the way, this is the undeniable right of the business owners. To maintain high profits, two types of cash flow need to be favourable: "cash in" has to be high and "cash out" needs to be low. Keep a healthy balance.

Once the general manager, who is usually responsible for the cash flow, signs a contract for a new build he has little influence on "cash in," but he has control over "cash out." He also carries the ultimate responsibility for the quality of products, as understood by the Quality Assurance concept, which according to Oxford Dictionaries is: "The maintenance of a desired level of quality in a service or product, especially by means of attention to every stage of the process of delivery or production." He plays a decisive role in keeping money in the company coffers.

At the same time, it is in the client's interest to cooperate with the yard so the business always has a positive cash flow. It isn't in his interest to starve the company of cash and bring the construction to a halt. So in this sense, maintaining a high profit and keeping the cost of quality low complements each other and is in the interest of both parties, the shipyard and the client. Keep the win-win scenario alive.

Something to remember:

The further you are from the problem,
the further you are from the profit

Quality

Quality is an extremely ambiguous term. Every client insists upon quality, but each has a different perspective on the concept. Quality, complexity and grade are not interrelated. A yacht can be highly complex and of low overall quality; conversely, it can be simple and of superb quality. "In many cases to 'qualify' quality in a contract, I choose a yacht known to the client and designer (preferably previously built by the yard) to pin down the 'quality' for contractual purposes," says Bill Sanderson of International Yacht Collection.

The cost of quality depends on whether it is assured or controlled. It also depends on the place it is managed from – at the top or on the floor level, top-down or bottom-up.

The term "cost of quality" refers to the costs that are incurred to prevent, detect and remove defects from products. It does not refer to expensive or very high-quality materials or construction methods. Quality costs are categorized into four main types:

- Prevention costs (the cheapest)
- Appraisal costs

- Internal failure costs, and
- External failure costs (the most expensive).

Something to remember:

"Quality starts in the boardroom" —Dr. W. Edwards Deming

Process Checking vs. Product Checking

Quality Assurance (QA) is the strategy of focusing on defect prevention; it manages the quality of inputs. Quality Control (QC) verifies the quality of the output and it is the strategy for detection, the process of eliminating existing defects.

Even the best Quality Management System (QMS) cannot withstand the demands of time. This is where a Six Sigma Plus process will play a role. Six Sigma Plus was developed to accelerate improvements of processes, products and services, and to reduce costs and improve quality.

Figure 12 - Process Checking vs. Product Checking

Keep Changes to a Minimum

Changes cost money and these costs are borne by the yard, the client, or both. In any case, both parties are affected, even if a change is fully paid for by the client. One change here and there to one boat will not make much difference, but if the yard has three 100-foot LOA boats under construction that means it produces 2 boats per year (assuming 18 months construction time per boat). Changes or additions to the 100-foot LOA sailing vessel may accrue to approximately $800,000 (if the product was well defined), which converted to man-hours results in approximately 16,000 hours (assume $100/h). This is about 20 workers, additionally occupied, for 20 weeks. This will have a significant effect on current projects and the construction schedule of the next projects. Somehow the yard has to endure and accommodate this impediment.

Changes in any project, and superyacht projects particularly, are unavoidable. While changes usually help to manage the scope of the project, they almost always negatively influence the cost and schedule. Only a professional approach to changes, such as a properly designed procedure of variation order (VO) (sometimes called a change order (CO) or change request (CR) may weaken their negative impact.

Variation Order (VO)

Variation orders (VO) are one of the most important documents used during the execution of a project because these orders impose changes to the highest document in the hierarchy of contract documents: the contract itself. This is why the procedure for a change should be clearly defined and be part of the contract. Such a procedure should include at least the following:

- Form (template) of variation order.
- Hourly rate and method of calculation.
- How a VO will affect project milestones.
- Under what circumstance the yard has the right to extend the construction schedule.
- Under what circumstance the yard may refuse to accept a request for change and continue the project.
- Who has the right to approve a VO.

For the yard, issuing a VO is a chance to gain additional funding. Don't be greedy – create a win-win environment. For those who prepare and approve variation orders, my advice is to give until it hurts, but no "gold plating," either! And if you offer to make a small, free concession, inform the client and make a note on a separate list. You may need to remind him about these "freebies" during the handover.

Advice for the client: Changes requested should be paid well and quickly, because "he gives twice who gives promptly".

The cost of any changes rises over time. So, the strategy here is that if you have to make changes, make them as close to the beginning of the project as possible.

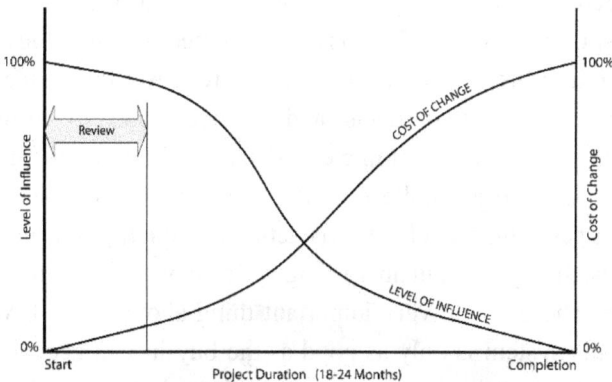

Figure 13 - Influence vs. Cost of Change

Modus Operandi

"If you don't have a formal methodology and process, you are relying on one manager to deliver your goals. That *ad hoc* approach creates a much higher probability of risk." —Ricardo Viana Vargas of Denmark

The Latin words "modus operandi", meaning method of operation, predominantly relate to criminal investigations where crimes with a similar pattern may prove to be committed by the same person. Similarly, a methodology leading to success can be used again and again, with confidence that it will lead to the same successes. Put a successful method in place for getting there and the results will have a tendency to follow on their own. My point is, you want to repeat a method of operation that is successful...don't reinvent the wheel every time.

The process of systemizing how things are done, the "way of operating", is known as a Business Management System (BMS) or Quality Management System (QMS), or Business Process Management (BPM), a proven way of keeping profit high and the cost of quality low. BPM is a management methodology used to communicate to employees what is required to produce the desired quality of products and services, and to influence employee actions to complete tasks according to the quality specifications. It is a collection of business processes focused on the best performance, client satisfaction and quality of product or services. It is all about increasing efficiency, effectiveness and quality. There is one very important thing about BMS, QMS or BPM: the system is only as good as the buy-in you get from the organization and senior management and how well it is implemented.

Of all the "regimes", the ISO 9000 series of standards is probably the most widely implemented worldwide. ISO international standards ensure that products and services are safe, reliable and of desired and repeatable quality. ISO 9000 series may not cover all necessary requirements of a business management system, but it is a good beginning.

A small boatyard may not have an official business management system; they just do things "their way". It may go like this: once upon a time, on a frosty morning in a small shed on the riverbank, a small crew is just about to start building a new wooden sloop. On the walls are hanging sets of wooden patterns representing shapes of frames. These are the patterns for various boats built in the past. "Skip, which of the boats are we going to build this time?" asks one worker. The owner of the yard looks around and without too much hesitation, still keeping his cigar in mouth, says, "Take five bow's patterns from this set, seven patterns from this one and we will complete the stern with a few of these." "No problem, Skip." A few months later the boat is launched and it is time to set the mast. "Skip, where should we step a mast?" Chewing his tobacco, he contemplates for a while. He spits his tobacco into the distance between the tenth and eleventh frame and says, "There!"

In larger boatyards the procedures are recorded to ensure everyone is clear on who does what and how. The beauty of a QMS is that it can be – and it should be – continuously improved upon. (See Lessons Learned.)

How important it is for a client that the yard has a QMS (Quality Management System), an ISO or any system that guides everyday activities is described in the following story. Three of us representing a new, emerging shipyard went to Atlanta to offer our services to a potential client. Our combined experience consisted of many years in various shipyards, participation in America's Cup, building superyachts, construction of Navy and cargo vessels, etc. Believe me, we tried to sell ourselves as best as we could. Everything was going fine. Then the client representative asked if we had already set up a procurement department in our new shipyard, and if we had any procedures for it. We were still in an early stage of development and although our preparations for a full-blown ISO system were very advanced, we had to answer that we did have a purchasing department, but we did not have any procurement policy, or procedures, or flow charts at this moment. After that answer, I felt that we lost the chance to impress the client and his entourage. The meeting had deflated. They suddenly lost confidence in us and felt dispirited. We went home empty-handed and they allocated the project to a different yard. C'est la vie! It was a great lesson and we realised that we had to finalize our ISO system as soon as possible.

Something to remember:

Magic is in the process.
Have a **SYSTEM**: Save-Your-Self-Time-Energy & Money

Culture eats leadership for breakfast...and for lunch.

According to Ellen Wallach of IBM, "Organization culture is like pornography; it is hard to define, but you know it when you see it."

Some insights on the topic of culture:

- One person is a personality, two a relationship, and three-plus people are the start of the creation of a culture.
- Each individual contributes positively or negatively to the culture.
- In a business where you deal directly with customers, your culture is your brand.

- First, culture forms by itself – you can't invent it. However, you can develop, enhance, and improve it.

Company culture is usually not very well understood and its value is underestimated, but these days it is recognised as more important than strategy. Culture is a manifestation of ideas, customs, and the social behavior of people or a society. Strategies you can modify by one stroke of a pen, but for changing a culture you may wait for years. According to Erling S. Andersen, Norwegian School of Management BI, company culture affects projects and services offered in either a positive or a negative way: "There is generally a great resemblance between the actual cultures of the base organizations and their projects. The culture that is first-ranked in the projects is nearly always first-ranked in the base organization as well."

An organization's culture describes the ways of thinking, behaviors and beliefs that an organization's members have in common. This is a collection of shared values, norms and expectations that govern the way people approach their work and interact with each other. The crux of the matter is that the culture (or character) of an organization, whether it is a company, a vessel, an army unit, etc., depends entirely on the head of the company, the CEO or the general manager. Robert Moore, author of *A Time to Die: The Kursk Disaster*, depicts this in his book: "Despite the presence of dozens of highly trained specialists on board a submarine, the character of the boat is defined by her commanding officer". More straightforward, this can be paraphrased to: "A fish rots from the head down". Please note that there is another book by the same author, similarly titled *A Time to Die: The Untold Story of the Kursk Tragedy*.

When Jack Ma started Alibaba on February 21, 1999, he asked seventeen friends to gather in his second-floor apartment at the

Lakeside Gardens in Hangzhou.

"Alibaba founder Jack Ma sees a strong corporate culture as critical to its success." —*Harvard Business Review.*

Marcus Goldman, the co-founder of Goldman Sachs Group, recalls those times. "Ma was able to imbue the new project with the same culture of his existing company, while keeping it totally separate. The decision to launch Taobao out of the apartment that spawned Alibaba worked so well that Ma took the same approach with Alipay, the company's digital payment business, which launched in 2004 also out of the apartment ... The whole place stank – all those instant noodles. Jack's ideas were not entirely original–- they had been tried in other countries. But he was completely dedicated to making them work in China. I was moved by what I saw."

He was adamant to repeat his first success by copying the culture.

Mark Stothard of Echo Yachts tries to create a culture based on the acronym **REACH** :

Respect:	earn it
Ego:	lose it
Agro:	don't need it
Care:	take it
Humble:	be it!

A Bad Attitude Is Like a Flat Tire...

The Merriam-Webster Dictionary explains attitude as "the way you think and feel about someone or something". Here is an attitude of Jerry Rice, regarded as the best wide receiver to ever play in the National (US) Football League: "Today I will do what others won't, so tomorrow I can accomplish what others can't." If each team member expressed such an approach, you could be sure of success. Otherwise, a bad attitude is like a flat tire: if you don't change it, you'll never go anywhere.

S.W.E.A.R.

S – Sail
W – While
E – Everybody (meaning: All)
A – Are
R – Resting

I developed this attitude for myself and my crew whilst racing on the Baltic and North Seas. The races were long enough to be at sea for a couple of nights (up to one week) but short enough to allow the deprivation of a bit of sleep. Most of the other crews usually maintained a normal 24-hour watch system. We had all hands on deck during the night, making sure we maintained agility and a competitive advantage at all times. This strategy worked well for us. Although building a superyacht and maintaining high alertness during a two-year project would be a bit demanding, a similar philosophy can be adopted: when everyone else's mind is dulled or distracted, stay doubly vigilant.

Something to remember:

"First of all, you've got to focus on getting the right people, of the right cultural fit, on to the right seats on the bus. Secondly, you've got to focus on trust, communication and delivery value to clients. The third step is to put in place the right place the right systems and processes around governance. If you get step one, two and three right, then profits and finance will look after themselves. They will naturally flow". —John Poulsen of Squire Sanders

It is unlikely that having scrupulously followed these strategies a project would derail. But it can happen. Then what? When it does happen, it's time to go into diagnosis mode.

STRATEGY ELEVEN

Fix it yourself or look for the "Man in Black"

WHEN ALL OTHER MEASURES FAIL

Two boat-builders chatting during smoko. One is a pessimist, and the other is an optimist. The first one (the pessimist) says, "Look, this project is not going very well. Quality is falling, we are well behind schedule, cost is over by 30 percent already, morale is low ... It can't be darker than this; it can't get any worse!" The second one (the optimist) replies, "Sure it can!"

So, you have done it all. You followed all these strategies and the project is still not going well. Ten months have elapsed building a yacht. Launching should be in twelve months, but everything indicates that it will be delayed and the yard is not able to predict how big the delay will be. Everybody can see that it will be significant. To make this matter worse, it looks like the cost will be exceeded and the quality of the performance is already below the expected standard. The client hired a well-known naval architect who supplied a set of drawings. The interior is also designed by a well-known interior designer. The client and his team have made a careful selection of yard and yet the project is not going satisfactorily. The yard design office is working on schematics, reticulation plans and systems on-board, etc.; a client representative has good expertise; and the future captain has considerable marine seniority and serves professional advice, but some "unseen forces" are causing the project to not be moving in the right direction.

Questions are asked: what's going on? The situation becomes tense. To salvage the project, the client hires independent specialists: first, he hires another naval architect, then a project management specialist, then a contract management expert and finally, the attorney(s). In connection with these moves the situation becomes even more strained and the shipyard takes a defensive position. More and more documents begin to walk from hand to hand (of course, with receipts). This tense situation only worsens the construction performance: morale is going down, causing an inferior quality of execution and increasing delays. Costs are rising!

In such situations the ego plays a big and devastating role. The human resistance to change slows down the rectifying process. NIH syndrome (Not Invented Here) paralyses progress. There is a struggle for its own interest and in the end comes the worst approach – to win at all costs.

You have three options

So, now you have three options:

- Option 1 – Review the documentation, processes and ability of your team, find out where the problems are, look for details, and fix it yourself.
- Option 2 – Look for "the man in black". Arbitration/mediation, Alternative Dispute Resolution (ADR).
- Option 3 – Go to litigation; take the dispute to court.

Regarding Option 1 – "God Is in the Details"

"God is in the details." —Ludwig Mies van der Rohe (1886-1969)

The most expensive hyphen in history

On July 22, 1962, NASA took on the first interplanetary mission with spacecraft *Mariner 1*. This mission was cut abruptly, 293 seconds into the mission. Those of you who have had the "privilege" to use old punch cards for computers will remember how easy it was to make a mistake. A simple error in NASA's code, a hyphen spelling error, caused the grounding of the *Mariner 1* mission, which cost $18.5 million. The spacecraft had to be destroyed because of a change in destination which sent the spacecraft on a collision path with Earth. The moral of this story? Check everything, double-check it, then check it again.

Of all the countries in the world, only three backwaters still use the archaic Imperial system of weights and measures. So, watch all those products coming to your project from Liberia, Myanmar and the United States of America. Although to be fair to the U.S.A., metric units are a standard in science, medicine, and government, including the U.S. Armed Forces, as well as some other sectors of U.S. industry. To be honest, I still love to use feet to describe the length of a yacht. Am I behind the times?

An old anecdote says: old carpenters prefer to use the Imperial system because it is easier for them to remember an integer number of inches plus a fraction than a measurement in millimetres. So this is your option: review the documentation, processes and team, find out where the problems are, look for details, and fix it yourself. This could be easier said than done because if you've found yourself in this situation, the question then becomes, what have you been doing so far? There is a good chance that you will commit the same mistakes again and blame the same people for the failures. Still, before you get into big trouble (like Option 2 or 3), do yourself a favour and start with this option first anyway. Review the entire project's documentation, design, communication, schedules, procedures, etc. and look for flaws. While doing this, mind the details. Even if you are not able to fix everything by yourself, going through this process you will be better prepared for the next stage (Option 2 or 3, if necessary).

Any business, including the business of building a superyacht, is like a yacht race. You do not win by doing extraordinary things; rather, you win by eliminating mistakes. In this project, probably somewhere along the way, a few mistakes have not been eliminated. Each person in the organization is required to pay attention to details, leaders particularly.

"Never neglect details. When everyone's mind is dulled or distracted the leader must be doubly vigilant. All great ideas and visions in the world are worthless if they can't be implemented rapidly and efficiently. Good leaders delegate and empower others liberally, but they pay attention to details, every day." — General Colin Powell

So, before you involve any third party (e.g. change agent, arbitrator or attorney), go through the various documents and establish what went wrong and what needs to be fixed urgently, in order to move forward. Cooperate like hell. The first thing to do is to read the contract. You may perform a cause-and-effect analysis or even several of them.

Regarding Option 2 – Look for the "Men in Black"

The "man in black" is neither change expert, nor attorney, nor any specific specialist. He could be an arbiter or mediator but not necessarily. He is just a well-known and all-around respected expert in the marine industry. He has been here and there. Does he have to be in a black suit? Of course not, but men in black are synonymous with professionalism and respect. Just kidding.

Who exactly is he?

- He has good interviewing skills.
- He has business analysis knowledge.
- He understands the complexity of the boat design process.
- He has very good knowledge of boat-building processes.
- He is a good project and contract manager.
- He is able to lead changes.
- He knows how to bring people to the table and mediate.

I would be surprised if neither the self-assessment nor "man in black" options work. But if not, here is your last resource:

Regarding Option 3 – Litigation

Involving attorneys could mean a costly and prolonged process. If you have to involve one or the whole legal team, get the best. This is how Malcolm Gladwell in his book *Outliers* describes good attorneys: "If good attorneys do not outsmart you, they will outwork you, and if they can't outwork you, they will win through sheer intimidation."

Leave it all to them. Suffer the consequences and pay the bills.

For all of the above options, you would probably have to introduce the change management process. Bringing in an independent change leader would make it all easier for you.

Change management – Herbal Treatment vs. Emergency Surgery

If it becomes apparent that a project is stalling due to some organisational, procedural or managerial deficiency, changes will be indispensable. Adopt a change management methodology (for instance PROSCI application). You may get a change agent/leader to help you to introduce these changes, but you need to know what you want to change. He will only facilitate changes.

According to the Change Management Institute, "Change management is an approach to shifting/transitioning individuals, teams and organizations from a current state to a desired future state."

There are two ways to introduce a change. You can work from the inside, from the bottom up, or you can come in from the outside and change things from the top down. In other words, you can introduce changes by evolution or revolution, or use an "herbal treatment" or perform "emergency surgery".

When you take the top-down approach, you'd better know damn well what you're doing, because it will be like running through a minefield. There will always be resistance to change.

The best way to change a system is to work through it as a bottom-up insider, quietly chipping away at standard operating procedures and creating small opportunities to do what you really want to do until you achieve real success. But time is running out. Will you have enough time to do it this way? Can you afford the "herbal treatment," or should you go straight for the "emergency surgery"? If you need to implement changes in a hurry, make sure that at least top players buy in.

Something to remember:

"You never change things by fighting the existing reality.
To change something, build a new model that makes
The existing model obsolete" —Buckminster Fuller

STRATEGY TWELVE

Don't damage this, by fixing that!

JUDGEMENT DAY

Launching. Finally it is time to put the vessel in the water. This is the moment everybody is waiting for. Launching has to be approached with the utmost attention. Many things can go wrong, as we have seen so many times. Expect the best, but be prepared for the worst. Particularly important are the last forty-eight hours before launching and launching itself. This process is dependent on so many details that careful, detailed planning of activities, sometimes to five-minute intervals, is a must. Each activity must be assigned to an individual, who will be responsible for its completion.

Sea Trials & Commissioning

After launching there are the sea trials and commissioning. Time spent on harbour and sea trials should not be negotiable. However, for a well-performing yard this should not take more than one month, maximum two months. A project should have an acceptance program as part of the contract. It is essential that no trials should take place without proper planning and clear lines of authority. Here is the list of the basic activities for sea trials and commissioning:

- Stepping the mast
- Checking performance of all systems
- Checking the noise level in each cabin and engine room
- Fuel consumption
- Speed mile (checking max velocity)
- Sailing
- Centre of gravity and stability
- Live on board

Note that throughout the trials – in fact, up until the handover – the vessel is in the hands of the yard's trials captain. It is a common courtesy to invite the future captain to be present on the bridge, and often the trials captain will offer the chance for the future captain to try manoeuvring the boat so he can do a better job when the owner finally steps aboard for his first cruise.

After launching and during the sea trials the owner would usually like to accommodate his crew on board. This is not recommended. They should live in a hotel until the handover. Why? Simply because the boat has not been handed over yet. During this last stage, there are still a lot of tradesmen, suppliers, consultants, photographers, etc. who move around, who have tools, who need space to efficiently finish their job. The boat is already very cramped. Imagine that you have an additional two or three people living on board. They bring their own stuff, they bring groceries, they cook, and they use the showers... It would unnecessarily add to the already existing havoc.

How Much Do We Damage This, By Fixing That?

Handing the superyacht over can be a stressful moment. It is like handing over your homework to a teacher and waiting for his comments. If the project went well, there is nothing to worry about. However, there will be always things to rectify– some vital, some less important. Now it is up to the owner's wisdom: "How much do we damage this, by fixing that"?

A yacht is seldom delivered 100 percent complete and ready with all systems working. It is just too complicated and it takes a lot of time to sort out all the problems on such a sophisticated unit. The yard will do its very best, and as the handover looms, the project manager will see the snag list (hopefully) get shorter and shorter, but as system are run in, problems are bound to crop up and sometimes delivery of replacement parts or incompatibilities make it impossible to have the boat running perfectly by the time the client wants to take his boat away. It is essential to have a plan in place that can accommodate this situation, which may include signing off on the snag list when the yacht is handed over, a return to the yard after the owner's first cruise, or a period made available elsewhere for a flying squad of yard workers to meet the vessel somewhere to correct any snags that remain. Practically, a yacht is generally at its absolute best two or three years after launching, after all the snags have been eliminated by the yacht's engineer and before anything has started to seriously wear out.

Celebration

This is the time to reflect on accomplishments. Was this project a success? Well, if you can look back and be satisfied and look forward and be excited, you can say yes, this project was a success.

Closing the Project: Lessons Learned

Sometimes, even a big, mature construction company, when closing a multimillion-dollar project that didn't go too well, fails to learn valuable lessons. Failing to do this means they are missing out on a great opportunity to draw conclusions and learn from their mistakes.

The purpose of collating all the experiences in one document is to capture any factors which have positively or negatively influenced the project. By reviewing the processes and performance of the project team you may gain additional knowledge, skills and experience – lessons learned. They should be identified, categorised and noted. This in return will hopefully be a catalyst to necessary adjustments in the methods (QMS) of the next projects to achieve an even better result. As the Chinese philosopher Confucius (551-479 BC) says, "Our greatest glory is not in ever falling, but in rising every time we fall".

Something to remember:

"The most important thing in life is not to capitalize on your successes - any fool can do that. The really important thing is to profit from your mistakes." —William Bolitho, from *Twelve Against the Gods*

STRATEGY THIRTEEN

Have fun!

"SORROW, COLD AND HUNGER ARE BAD BUSINESS PARTNERS"

From the book of *The Black Obelisk* by Erich Maria Remarque.

Finally, it is time to touch on the relationship between the future owners of the superyacht and the shipyard representatives. In my experience, from the earliest stages of the negotiation, preparation of contracts, and discussions about the project to the final handover, despite some difficult conversations and hard decisions, it is all accomplished in a great atmosphere with excellent food and wine. Very often, even after long debates, there is still no conclusion – honestly held views (opinions) have been exchanged, but positions are not agreed upon. Despite this, sessions like that, no matter how tough, are finished with either a good lunch or dinner. This is the glamorous part of building superyachts and definitely helps to smooth the sailing through sometimes rough waters. Such a spirit of cooperation should be maintained till the end of the project and beyond.

The following story has nothing to do with strategies, but it has something to do with the win-win scenario. We went to London to finalise the contract for a fast powerboat. After a whole day of intensive work on specification and cost adjusting, the client invited all involved and his family to a restaurant for dinner. We sat at a round table. On my right was the client representative and on my left, his wife. Behind her was my boss, owner of the yard, then the client's whole family, him, his wife and his two kids. Dinner was going nicely with a cooperative atmosphere. It was the type of restaurant where dinner is served in many, but very small, portions; for instance, a fried grass. The conversation swung from one subject to another, and we hardly talked about the future project although "sailing stories from the cockpit" were very popular. We really enjoyed a relaxing evening after a long and hard day working.

All of a sudden, everybody's eyes turned to the lady on my left, so I looked at her too. She was red, her eyes had popped out, she was out of breath... she was choking. I knew I had to do something and I knew I had to be quick and decisive. At this moment I completely forgot the method used by Robin Williams on Pierce Brosnan in his famous movie *Mrs. Doubtfire*. My actions had become instinctive. So, I did what we always used to do at home in this situation when I was young. With the whole might of my left hand and the vigor corresponding to the moment, I whacked her in her back.

Well, she was a woman of a slim and fragile figure. I instantly realised that I had probably overdone it a bit. Everybody looked concerned and immediately changed the direction of their gaze from her to me. The expression on their faces changed from concerned to puzzled then to concerned again. However, I wasn't sure of the result of my first "helping hand", so I hit her again. It got very quiet. The revellers were dumbfounded.

The expression on my boss's face was telling me, "Tomek, stop these barbarian practices, otherwise we will lose the contract." I became very concerned. Was I helping this poor lady or had I made her worse? Lucky for me, she started breathing, although still with some difficulties (not surprising, since after a blow like that, anybody would have a breathing problem). Her eyes reverted to their normal position and her skin colour started returning. After a while everything returned to normal. We continued our enjoyable dinner. In the next couple of days we completed negotiations and went home with a signed contract. When construction started, however, and the client representative and his wife came for the first inspection, I noticed that whenever I raised my hand to show something or correct my hair, she gently eluded me. I don't blame her. What was important at the end was a win-win outcome: she survived, and we won the contract.

Figure 14 – 56m LOA Ketch, Glowacki Design

*   *   *

Tomek M. Glowacki

FINAL NOTES

Don't forget about technical matters

Many yacht-yards would save themselves trouble and avoid embarrassment by taking better care of just only the following three technical matters:

CENTRE OF GRAVITY

The stern and the centre of gravity are often too high, sometimes so high that boat flips belly-up straight after launching.

Although the designer tries to be as precise as possible, he probably will not admit that at the end he adds about 15 to 20 percent to the weights that somehow get omitted from the calculation. It is necessary to keep monitoring all of the weights going into the boat during construction. Follow up with calculations and occasionally weigh the boat to find out what the actual weight is. Unfortunately, the actual position of the centre of gravity can only be found by doing an inclining test when the boat is in the water, but precise calculations can give us a pretty good indication.

FAIRING AND PAINTING

Everyone seems to have his or her own definition or vision of the desired level of finish. In fact, even during the course of a paint job owners' opinions and standards can change. So what started out appearing great can sometimes be recognised as the job progresses and the finish becomes more heavily scrutinised.

The owner usually has a picture in his mind: "Make it shiny like my new BMW and I will be rapt!" However, when the day arrives, it is not the mirror-finish he'd anticipated since day one. The gloss on his boat is in fact higher in absolute terms than on his car. In both cases orange peel is present; however the peel is of a finer and more even texture on the boat. This finer peel lowers the DOI (Distinction of Image) at 15^0 on the boat but improves the DOI at 60^0, whereas the coarser peel lowers the DOI on the car at 60^0 but increases the DOI at 15^0.

Who's right, the owner or the painter? Neither, actually, because no meaningful values were put to the owner's expectations or the painter's capabilities. Prior to the job starting, the painting contractor must be given the opportunity to demonstrate he/she can achieve the required finish quality to the ultimate decision-maker (most often the owner).This can only be in everyone's best interest! The owner has no argument if he finds "unacceptable" faults. The painter knows beforehand the finish required and can price it accordingly. The boat builder has mitigated any likelihood of conflict. Everyone has the confidence that the products, tradespeople and facilities are up to providing the agreed-upon standard of excellent work.

VIBRATION AND NOISE CONTROL

In a short space of time, I have seen on LinkedIn several cries for help because after launching a vessel, vibration and/or noise level in the cabins was unacceptable. What was even more unsettling in this case is how many "experts" - volunteers rushed in with advice without even knowing exactly what and where the problem was. These problems can be fixed, but is much harder to do it when vessel is completed and it could be a costly exercise, while during the design stage it may only take couple of short strokes of a pen. Here is an easy solution: engage an engineering specialist who provides services relevant to noise and vibration control at the early stage of design.

CONCLUSION

What I have presented in this book is only the basics, touching on and overlapping the main disciplines involved in superyacht construction such as project management, contract management, naval architecture, design, leadership, stakeholder management, risk involved, etc. There are many ways to enhance the performance of a superyacht project. It definitely helps to have good discipline and an open mind to new skills and knowledge.

But I have to warn you – the more you learn, the more you will realise how little you know, and the more you would like to know. Isaac Asimov (1920-1992) said this: "the saddest aspect of life is that science gathers knowledge faster than society gathers wisdom". And, the late futurist Buckminster Fuller once observed that the conventional education system is at least half a century behind what science currently knows. To a large extent, that remains true today, and because of this, we all need constant education and accept CHANGE.

If you want to succeed in project management of a superyacht project, you must gain specific knowledge and skills, and know your destination and direction at all times. Prepare winning strategies, stick to your plans and look out for the unexpected.

Something to Remember:

The ability to lead is very important. But, can you lead if you don't know how to manage?

* * *

ABOUT THE AUTHOR

Tomasz (Tomek) Michal Glowacki (pronounce: Gwovatzki) was born and brought up in Poland. An engineer, entrepreneur, inventor, yacht designer, certified Project Manager Professional, Six Sigma Black Belt in continuous improvement and Ocean-Going Yacht-Master, he has forty years of experience in a diverse range of industries in roles ranging from designer through project manager to general manager and member of boards of directors. His experience comes from projects ranging from several thousands to 450 million dollars. He led the start-up of a superyacht facility in Whangarei, New Zealand, as well as assessments and improvement of shipyards worldwide. He is an independent management consultant specialising in business analysis, project management, continuous improvement, culture change, business turnaround, and strategy execution. His passions are naval architecture, biographies, WWII history, sailing, cycling and skiing. He cares about the natural environment and animal welfare.

For conferences, workshops and seminars presentations see:

tomek@tegassociates.co.nz
www.glowackimaritime.co

www.ingramcontent.com/pod-product-compliance
Lightning Source LLC
Chambersburg PA
CBHW061257220326
41599CB00028B/5686